BURIED ALIVE

BURIED ALIVE

THE TRUE STORY OF THE
CHILEAN MINING DISASTER
AND THE EXTRAORDINARY
RESCUE AT CAMP HOPE

MANUEL PINO TORO

FOREWORD BY NATALIE MORALES

palgrave
macmillan

FEB
2012
m(49)

To Luis Alberto Pino Bustos, proudly looking down from heaven above.

To my parents and family, for their unconditional love.

To Loreto, for her patience, love and unwavering support.

ACKNOWLEDGMENTS

Special thanks to my agent Diane Stockwell. Also thanks to Andrea Montejo; at Palgrave Macmillan, Airié Stuart, Luba Ostashevsky, Donna Cherry, and Laura Lancaster; and to my translator, Michael Carr.

CONTENTS

FOREWORD

Natalie Morales

I t was the story that captivated the world as millions of people—by some accounts even a billion—watched to see each and every one of Los 33 be pulled from the depths of their underground hell and brought back to life. What unfolded in the sixty-nine days leading up to the rescue was portrayed as a story of unity and courage—celebrating the heroes, the thirty-three miners who survived the impossible, as well as the accomplishments of a nation that pulled off the greatest rescue of all time. We all know the ending, of course, but we are just as intrigued to know what really was going on with each passing day. How did they survive those first seventeen days with no communication and no additional food or water? Who are Los 33? What is the Chilean government not sharing—and why? Is there more to the rescue story that has yet to be told?

We may never have a complete account or true understanding of all that was happening in the Atacama desert at the San José Mine, but *Buried Alive,* by celebrated Chilean journalist Manuel Pino Toro, gives us a complete and accurate account of how the story unfolded and just how each of the thirty-three miners played a crucial role in their survival and rescue. Toro reveals how the miracle happened and takes the reader on a journey of amazing discovery and high drama. Fortunately, I too was among the nearly two thousand journalists who were there to witness history; otherwise I'm not sure I would truly believe it all. It's a story we simply can't get enough of, and Toro has a unique perspective as one of Chile's foremost journalists.

INTRODUCTION

Six hundred miles south of Santiago, Chile, lies the Lota coal mine—five thousand meters of tunnels submerged beneath the Pacific Ocean. I was there to write a news article shortly before February 27, 2010, when the area was devastated by the tenth largest earthquake ever recorded: 8.5 on the Richter scale.

I was inside the tunnel with the miners when it happened. That experience beneath the sea struck me: the suffocating heat, the fatigued faces of the miners toiling away in a space so tight they had to move about almost on all fours. These men's lives have been much written about. But even the most thorough descriptions of these workers, who set out from their modest homes each night to enter another sort of darkness—that of the mine, where they pound away at the ore until it cracks loose and is hauled away—are incomplete.

Much has been written about mining in the north of Chile, a zone rich in nitrates and copper. Its history began in Chañarcillo with the extraction of silver, the mineral that made this the first place in South America with a railroad to transport material, provisions, and passengers. But despite the abundance of chronicles written about these exploits, history now requires a new narrative—one that, if it were not true, would sound most unlikely.

Almost 500 miles north of Santiago, in the middle of the Atacama Desert, just after midday on August 5, 2010, time paused, and a new story began about men's labors underground. It started with just another cave-in beneath the Chilean pampa. This quickly became a mass-media event that flashed across all national borders and ended up being broadcast live by all the major television networks to millions of people throughout the world.

A gigantic rock had fallen, blocking the main access to the San José mine near the city of Copiapó, leaving thirty-three miners trapped for sixty-eight days.

The accident brought to light yet again the lack of safety inside the mines. For me, it felt like a return to Lota, where the coal now costs too much to extract, or, northward, to Chañarcillo, where first the silver played out and then the nitrates followed—and, with these, the dreams of generations.

Now all hopes feed on copper, the latest life buoy for many families, including those of the thirty-three workers who were buried in the San José mine.

In Lota, the miners left, but up north in the desert, it was not the will of man but the caprice of nature that ground the mining work to a halt and put the lives of so many in doubt.

A few weeks earlier, I was waiting to get my passport stamped on reentering Chile at the Santiago airport. The immigration officer asked for my help communicating with an Asian youth, as if I could somehow help the official understand what he was saying. Unable to speak any language but his own, the youth could not state his purpose for visiting the country.

Defeated by the situation, the woman stamped his passport, saying, "I've given him thirty days. They're all Chinese; they come to work as miners." Thanking me for my vain effort at interpreting, she gave a shrug and welcomed me into the country.

I left and followed the young Asian man but quickly lost him when he didn't wait to pick up any baggage. I had to wonder, was the country importing workers . . . or perhaps slaves?

Hearing the news of the San José mine, I glued myself to the radio and remembered Lota. It wasn't hard to imagine what it might be like in the mine. I could even imagine that the men might have survived the collapse. For the miners I had interviewed deep beneath the sea at Lota seemed strong enough to bear all manner of misfortunes. I searched within for some reasonable way to tell myself, *Yes, they are alive,* but, of course, rationality pointed me to a different conclusion.

The giant block of stone that had broken free could not be removed from the mine's entrance. It was winter, and the desert was cold. The climate there is regulated by the Humboldt Current, which crosses the Pacific and blankets the Chilean coast each morning with a thick mist known as *camanchaca*. In the evening, you can see a deep blue sky and at night, the ends of the universe itself. This is why there are many astronomical observatories in the Atacama Desert.

Nowadays, though, the most advanced technology not only points upward, it also looks down into the depths of the earth. And the governments of ten countries were scrambling to put their best equipment and expertise into the effort to rescue the thirty-three miners of Atacama. I was struck by the response from all sectors of Chilean society toward some trapped workers in an industry that affords them little protection for the hard and dangerous work they do.

We see the familiar ritual of tragedy performed again and again, yet no one ever seems to take responsibility for it. Large companies who fail to comply with safety regulations are vilified in the press. A great hue and cry goes up, and those responsible are pursued with much fanfare, but the trials end, with few exceptions, in light sentences and endless lawsuits. So the biggest tragedy is what comes *after* the event. Wives, children, parents are left heartbroken. The mining companies express regret at the situation but are in no way obligated to compensate the families. The press might make noise, but regulators are bribed to look the other way and the cycle of neglect continues.

When unsafe and poorly regulated mines collapse, death is almost unavoidable. Six years earlier, in Río Turbo, which belongs to the Argentine part of Patagonia, some of the Chilean workers who daily crossed the border to scrape a living in a mine there became trapped. During coal extraction, the gas known as firedamp, produced by the mixing of methane with air, accumulates. If enough builds up, an explosion is all but inevitable. When the entrance is sealed, death by asphyxiation is certain. Thirty-seven men escaped that collapse, but fourteen, three of them Chilean, were trapped at nearly two thousand feet down, almost four and a half miles from the mouth of the mine. They all died. But, I told myself, the San José mine was about copper, not coal, and this meant hope, since the explosive component was not present.

I said to myself, *Pray there isn't an earthquake.* The year 2010 had been Chile's most seismically intense in decades. The north had suffered two quakes in a little less than ten years. Copiapó had escaped major cataclysms for a century—I hoped it could do so a little longer.

It would seem that Chile, a nation of individualists with little solidarity, was able to break through its well-defined social barriers to cope with the earthquake emergency. And this movement toward solidarity was new. None of the previous mining tragedies in the mines was able to bring it about.

But that first day, this was just a collapse. Yet the media opened the matter to speculation, which prompted many in their audience to unplug and turn off. There was no new information, and all the pointless repetition took interest away from the actual event. And when night fell in the capital, about 530 miles south of the San José mine, still no one seemed to know anything new. So I decided to go and see for myself.

Not forty-eight hours after the collapse, I found myself lying awake in the desert, wondering, *Perhaps they got out.* I would return to the site again and again with the same question, picking the brains of experts and authorities more knowledgeable than I. The rescue operation was just beginning, but even then I knew that it would be no easy thing to drill through nearly half a mile of rock and pluck a needle from a labyrinth.

In this story, there are those who brought hope, those who stayed alive by the promise of hope, and those who were victims of its ebb and flow. And as the participants struggled, humankind worked together and overcame its limitations and its differences— at least for a time—united in the greatest of all causes: saving lives.

CHAPTER ONE

THE ILL-
FATED SHIFT

Ximena Fuentealba shredded a piece of chicken and arranged the pieces lengthwise on a fresh *pan batido* she had just brought, nice and crispy, from the bakery. *Pan batido* is the name in several provinces of Chile, especially in the north, for a white-flour roll, baked with its surface split and ready to break open and eat in two halves. Called *marraqueta* in Santiago, it is a staple food on the Chilean menu. If, soon after a child is born, the family begins to do well, they say, "The baby arrived with the *marraqueta* under its arm."

With great care, Ximena carefully made each sandwich, one after another, stacking them in dozens, which she would bring as part of the meal for the workers in the San José mine. She wrapped them in white paper napkins and stacked them almost without looking, for she knew from long experience in the art of cooking that she had each one just right. As usual, at one P.M. she had the workers' lunches all ready.

Standing in the kitchen, just a few yards from the mine, Ximena glanced at her watch and stepped up her pace. At two, the day shift got off—her spoiled brats, who devoured everything she made for them. "Ximenita," they said, "you have a 'nun's touch' with cooking . . ." She heard the words often, and they filled her with pride. And why shouldn't they? It was a well-known fact that nuns were gifted at baking. She couldn't fail her boys.

Today, August 5, twenty minutes remained before the bell signaled the shift change, and everything was just the same as always.

But suddenly, the kitchen—built in a shipping container— began to shudder. A strange, powerful jostling of the earth, followed by a deafening crash, made Ximena gasp.

A *tremor,* she thought at once—after all, Chile was seismically active, and the country had just dug out of the February 27, 2010 earthquake, which devastated the southern region.

The sound bounced in all directions, and she felt as if the cavernous booming were too close, like something from a horror movie involving demonic possession. It was as if the earth beside her were writhing inside its own skin. The cracking, grinding noise wouldn't stop.

Ximena dropped the kitchen knife on the table and strode outside to see what the hell was going on. A thick cloud of dust had billowed up from the mouth of the mine. It looked like the scene of a disaster, bringing to memory images of the bombardment of the presidential palace back on September 11, 1973.

The passage of time seemed mangled and uncertain—there was simply too much going on. The mining equipment operators, like Ximena, were running back and forth. The sirens blared without a break, and panic overran the barren site out in the middle of the Atacama Desert, a few miles from the northern city of Copiapó.

"It's a *planchón!* . . . A *planchón!*" she shouted in desperation, using the miners' jargon meaning "snowfield"—a collapse inside a mine shaft or tunnel.

After the roaring crash, nothing. From inside the mine, only dust and silence; from this side, sirens and shouting and chaos. No one knew whether the workers on the shift managed to escape or lay entombed in a hell world of broken stone.

Overcome by anxiety and desperation, Ximena stood frozen in place. She reached up to smooth her hair, as if that ordinary movement might help dissipate her uneasiness. She steeled herself, gathered her wits, and prepared to render aid. Behind her, in the lunchroom, lay a box piled with sandwiches that would never be eaten.

The workers who had gathered on the surface were dashing off in all directions, bumping into each other and stumbling over their own feet, yelling, not knowing which way to go, while dust floated above them, as if mocking their useless efforts.

Now Ximena began to fear the worst. The crash held her paralyzed beside the shipping-container lunchroom. With her hair now in place, her hands flew to her mouth, as if to keep her heart from jumping out of her throat. She held out hope, but the silence from inside the mine filled her with dismay. Her hands went to her forehead, as if beseeching God and the Virgin. The seconds ticked by and became minutes, and from the other side there was still only silence.

Overwhelmed, she thought, *The only thing for certain is that something terrible has happened.*

In Copiapó, the horrible news spread like flood waters, starting with the mine's own people. The owners tried to hold the news back, but as usually happens, the truth proved irrepressible.

HURRY . . . HURRY

Worry grew and built on itself and multiplied among the Atacama's inhabitants. Conversations traveled in a thousand directions, repeated by those just learning of the accident—an accident that almost nothing was known about.

"The shit hit the fan, Papá. The *niños* didn't get out."

José Vega grew very still. "What's that, son?"

"The mine collapsed, and they don't know how many were in there. Only a miracle can save them."

"But . . . who told you?"

"It's on the radio."

Like Ximena, José Vega could not—did not want to—believe the awful news he had just heard in the flat, dry voice of his son Jonathan, telling him that his other son Alex was trapped in a collapsed mine. It hit him like an airplane hitting a mountainside, and his face showed it. Shaken and queasy, he silently mouthed the words that kept repeating over and over in his mind: *It's terrible, it's the worst . . .*

With a force that he seemed not to notice, he held the telephone clamped against his right ear, hearing his own pulse as he felt his blood pressure ramp up. In the sepulchral silence, he bit his nails in anger, in dread, in a flood of anguish.

At the other end of the line, Jonathan's voice rose with anxiety. "Papá, Papá . . . talk to me!"

Without replying, José Vega hung up the phone. The seasoned miner, seventy years old, was alone in his house in Arturo Prat de Copiapó, a town named after Chile's great naval hero. The rest of the family was out and about, running the usual errands on what should have been just another cold winter's day in early August.

The telephone began to ring insistently, but José was mired in a lethargy that made even standing up difficult. He felt impassive, powerless. In his mind he had traveled to the mine, which by happenstance bore his name, to join the crew of miners who, rumor had it, were trapped in the bottom of the shaft.

The telephone kept ringing until Jonathan managed to make his father hear him. "What's wrong, Papá, why don't you answer?"

José's reply turned into an order that Jonathon would have to obey.

"Son, go ask your bosses to give us permission to go to the mine immediately and see what's happened to your brother. Now, scoot!" he barked, leaving no room for discussion.

"We'll go right away, Papá."

"Go. Just do it. I'm going to work out a rescue plan. We'll go see what's going on with the boys." He called them "*niños,*" an affectionate term meaning "little boys." "We have to do something," he murmured. "And do it fast."

José Vega hung up the phone and hurried to the room in his house where he kept the tools that could help him with what he needed to do: save "*los niños.*"

While all this was happening in the north of the country, down in Santiago radio and television programs reported in disordered, fragmentary snippets on happenings they knew almost nothing about. All the audience could gather was that just now a terrible accident had befallen a mine in the area surrounding Copiapó.

The news announcers kept repeating the same thing, stalling for time while their reporters scrambled to get on location. For the moment, vague and redundant information was all they had to fill the airwaves with, with everyone promising to comunicate something more precise and substantial, more coherent, more *real,* than they had so far.

All was confusion, and the news channels turned their attention to this "breaking story," as the anchors call it when they know almost nothing of what they're trying to talk about. Even hours later, no one could explain what had happened, let alone the possible consequences of the tragic event. Nothing was known in Santiago—and little more in the place where the collapse occurred.

"IT'S ABOUT BRINGING HOME THE BACON"

Cave-ins were nothing new in the San José mine. The site had been in operation for more than 125 years—since 1881, according to the highway sign erected at the entrance—and its seemingly endless access ramp spiraled down almost a half mile below the surface.

The work was intense and complex, and the older miners had complained many times that they were removing the copper and gold ore even from the very columns supporting the tons of rock above their heads. The conditions in the mines were so perilous that every moment of their day posed a risk.

The place was famous throughout the region for being poorly designed and for not taking into account the risk to the miners. But there was also another undeniable fact: this was a source of jobs—something not to be taken for granted in an area where mining was essentially the only work available and where few knew any other trade.

The ore deposits in the San José mine were complicated, tricky to get at. Its operators knew this; still, it was a good economic opportunity for some—tough, seasoned men accustomed to the perilous trade they plied deep undergound. And after all, as Chileans said, *"Hay que parar la olla"*—one must work and bring home the bacon.

Fire engines and an ambulance arrived at the site. Ximena Fuentealba asked one of the foremen, a man in a black leather jacket and black boots, holding a hard hat in his left hand, if he thought the miners were still alive.

"We don't know," he replied in a stern voice, without looking at her.

Out of the blue, Ximena remembered her sandwiches. They were piled on the table awaiting diners who, by all indications, weren't coming this time. The confusion of feelings pent up inside her finally let go, and she could no longer hold back the tears or calm her erratic breathing.

A couple of hours went by, and relatives of the trapped miners began to show up. They had gotten word by cell phone from the workers on the next shift, who had come in after the accident and watched the drama unfold from the surface. They were friends and acquaintances of those who remained below.

So far, no one had a clear idea of the magnitude of the event. Speculations and rumors grew, contradicting each other and sowing yet more confusion among the families and the rest of Chile, who now knew of the disaster and were attentively following the fast-traveling news.

The Chilean desert is a vast expanse full of huge mines and also many smaller excavations, and rockfalls were common. But only rarely were miners trapped. The family members who arrived at the San José site expected the rescue to be quick and that they would be able to take their loved ones home right away. Their faith was fragile, but it was still there. They didn't imagine, nor want to accept, that the entire work crew had been swallowed up underground and that this time the wait would be much longer than anyone could conceive.

"No one came out," the managers of the mine informed them. They checked and rechecked the roster of workers. There were thirty-three in the mine—too many.

The order came: "We go in carefully, because the mine's still unstable and is settling." This made it dangerous, but they had to see at what point the giant spiral staircase that followed the inside of the hill had been cut off.

"Does anyone know if they've restocked the provisions in the refuge chamber?" Getting no answer, the question started making the rounds. In case a rescue should take longer than a day, it was necessary to think about what was available for the trapped men to eat and drink.

The young man responsible for the job crossed himself and gave a prayer of thanks. He could scarcely believe his intuition. That very morning, after several weeks of seeing the empty emergency cache at the bottom of the shaft, he had finally restocked the water and brought down cans of tuna and crackers. There was food below to last at least seventy-two hours. *Gracias, Dios,* he thought, crediting his Maker with the foresight.

On the pampa the hours slogged by, the people's anxiety mounting all the while, and no one had yet heard from the owners of the mine. They were not where they ought to be: here, at the scene of the tragedy. "They have to take charge of the rescue," people were saying. But their absence during the six long hours that had elapsed only served to heighten the confusion and uncertainty that were taking hold of everyone.

The new arrivals sat down on the rocks. Others wandered about the footpaths lightly worn into the hard ground. Most of the family members had never been to the mine, because their breadwinners didn't want to bring them to such an inhospitable place.

But now they could finally see for themselves what it was like to work here in the middle of the driest desert on earth, surrounded only by dunes and barrenness. Now they understood the daily sacrifice these men made to provide for their loved ones—these men from whom, even now at sunset, no one had yet heard a peep.

A PLEA TO GOD

It was late, and the bus arrived, as it did every day, to carry the workers back to Copiapó. This time, no one but Ximena got on. She sat down and spoke with the driver, her words tumbling out in little bursts from nervousness and the bumps in the road. He listened impassively, nodding yes and looking out his window, avoiding the reality.

"It's terrible, just terrible, what's happening," Ximena said. She hated to leave without knowing, but she was confident that tomorrow it would all be sorted out and she would again see the miners back on the surface. This was her prayer to God as the bus ate up mile after mile of dusty road.

Two hours later, back at her house and desperate for news, she opened the door and immediately flicked on the television. The place where she worked was on every channel. Some commentators, better informed now, were calling it the worst mining disaster in Chilean history. Shaken, Ximena turned off the TV. She didn't want to see, but she had a feeling the truth was much worse than they imagined.

"IT'S TOO MANY"

Now it was night—an awful night. News of the collapse continued to spread like wildfire. President Sebastián Piñera, only five months in office, got a call from Interior Minister Rodrigo Hinzpeter. President Piñera had just arrived in Ecuador, the first stop on a tour that was to culminate in his attendance at the inauguration ceremony of President Juan Manuel Santos in Colombia. According to witnesses at the scene, Piñera pressed his hand against the telephone, speechless for a moment, and buried his fingers under the collar of his white shirt in a nervous mannerism he hadn't displayed in months.

The number of those trapped—thirty-three—concerned him. "It's too many," the president said to Hinzpeter. Immediately, he instructed the minister of mining, Laurence Golborne, who had come with him to sign trade agreements with the Ecuadorian government, to streamline the signing process for tomorrow morning, return to Chile, and go to the site of the collapse.

Golborne had gotten word at the same time, in a text message from his undersecretary, Pablo Wagner. He ordered Wagner to get to the mine at once and send him a detailed report, as soon as possible, on the condition of the buried miners.

Then Golborne took a flight from Quito, Ecuador, via Lima, Peru, to Santiago. Once in the Chilean capital, he headed straight to Group 10 of the Chilean Air Force, a military installation located next to the Pudahuel International Airport, and took off in a plane for Copiapó. He arrived at three in the morning. An hour later, the minister was at the mine.

As dawn broke, Minister of Labor Camila Merino was coordinating the operations. She had broad experience in the mining sector, having worked for twelve years as a manager of the Soquimich corporation (Sociedad Química y Minera de Chile). All her efforts were focused on the ventilation shaft in the hill—a passageway that, if it was clear, would permit the slow evacuation of the trapped men. Early that morning, Merino issued the first statement by the government.

"There are two entrances to the mine," she said. "The ramp, which is eight meters wide and through which trucks can travel, is not a possibility, because it is totally collapsed. The solution is the ventilation shaft. This duct is clear, but we have to work carefully because if we cause another collapse, we put at risk not only the rescuers but also the possibility of getting the people out quickly. If the ventilation shaft collapses, we have problems."

The operation moved stealthily. The rescuers could not afford any errors born of haste, since the least mistake could trigger a major cave-in. No one knew yet whether the ventilation shaft was still functioning or whether it would even work as an exit route.

Nor was it clear whether the thirty-three had ever made it to the refuge chamber, supposedly the safest spot in the mine—or exactly what provisions they might find there.

"The information we have is that the refuge chamber is equipped for seventy-two hours. We hope that the people are at that spot, but so far we have not confirmed this. We are doing everything we can to solve this as quickly as possible," Minister

Merino told the journalists who had arrived on the site. Several of them were the main anchors of the most important television networks in the country. They had left their cozy offices with their well-staffed makeup stations to grab a spot wherever they could in the dusty area surrounding the mine, "to report from the site of the tragedy." Now they, too, were part of the story.

The governor of Atacama, Ximena Matas, was a lively blonde woman, fashionably attired in a bright red jacket. As the ranking regional authority, she arrived in a rush to the scene of disaster. Born and raised in Calama, another mining city, she had known the area since childhood, when her father provided transport services to the Chuquicamata mine. The governor made it clear that all their hopes were pinned on the refuge: "We hope they are in a part where there hasn't been a cave-in—and where, moreover, there is a place outfitted especially for emergencies. This refuge has air, food, and water," she announced with supreme confidence—or, at at any rate, with the confidence she wished to convey.

Wearing a hard hat, Matas spoke to the miners' family members, who now made up a sizable group. "These mines are entered via a ramp," she explained, "which is like a road going down, around and around in a spiral, and there's a part where the cave-in is. That's why no one can get past it: because there are tons of rock in the way."

Around her, the emergency plan was unfolding. One hundred thirty rescuers were hard at work. There were pickup trucks assigned to rescue and communications functions, and a specialized team of mining experts, out of the Michilla site, located north of Antofagasta.

Among the volunteers was a well-known, experienced rescuer from the area and José Vega, who had managed to get through the security measures around the perimeter of the site and bring in his equipment and tools to search for one of his sons, Alex Vega.

CHAPTER TWO

RECKLESS ACTS

Like his son Alex, José Vega was also a miner. After hearing about the mine collapse, he had come all the way from Copiapó and joined a group of rescue workers preparing to check out the ventilation shafts, which, for the moment, looked like the only way to free the miners caught below.

Quickly he exchanged opinions with the rest of the men, who, like himself, didn't know for certain the actual magnitude of the collapse. Still, his years of experience working small mines told him that the accident was indeed serious.

José went to work setting up safety equipment with the other rescuers. His large family, who had come with him, joined the relatives of the other trapped men. Little by little, families had been arriving at the site and were now demanding information. A large shapeless crowd was forming here in the middle of the desert—a group of people who had no real idea of the daily activities that went on here.

Vega was anxious to get inside the mine. However, he would have to wait a long while until he could relieve another team of rescuers, led by the distinguished firefighter Pedro Rivero, who for the past two hours had been searching for a way to get down to the trapped men.

At the moment, Rivero was hard at work deep inside the mine. The intense heat and confined space in the shaft he was trying to examine didn't prevent him from moving with a certain ease. He descended down the shaft tied to a 967-foot rope. At a certain point, he stopped. Peering into the darkness, he tried to visualize his surroundings.

He needed to know whether cave-ins were still happening inside the shaft. That information was vital to evaluating the safety conditions before pressing on with the search.

He continued slowly downward—an even more reckless act. Rivero asked his companions suspending him from the surface to stretch out more rope. He descended another couple of yards and stopped. In a better position now, he beheld a terrible sight: he could make out a huge wall before him, gigantic and imposing. This was the rock that broke off. There was no way. It was impossible to go any farther.

Stunned at the sight, he said, "This is a total collapse, not a simple cave-in. There is no way out."

Almost forty-eight hours since the accident, it was the talk of the entire world. Bit by bit, the family members gathered at the site to learn each other's stories. Lilianet Ramírez's husband, Mario Gómez, was among the trapped. Mario, age sixty-three, was one of the oldest men on the shift. Happy and full of life, he was always the first on the dance floor at parties and was known for his sunny disposition. Anyone would have thought him years younger than his actual age.

He had said he didn't trust the safety in the San José mine. He always told her they were having cave-ins, and right there, only last week, a young worker named Gino Cortez had gotten his leg cut off. What's more, she had a nephew who had been disabled by an accident in the mine.

Lilianet was trying to cheer up the younger women when someone interrupted and asked her to come to a little white tent pitched amid the rocks. It was the emergency team, made up of a psychiatrist and three psychologists, along with four social workers. She felt comforted when they told her their work was to help the family members endure the waiting, but it made her uneasy when they said they were also there to prepare everyone for the possibility of an unhappy outcome.

In the regional hospital there was also a special sitting room for family members and a cell phone to inform them about the situation of the trapped men. Lilianet's thoughts ran back and forth between confidence that the right things were being done, and the knowledge that no one could guarantee anything. Her anxiety warred with her faith that everything would turn out well.

With all the national media awaiting information, the regional government finally issued a communiqué.

"As the Emergency Operations Committee of the Regional Government, we are bringing to bear every action, human resource, and material to ease the effects of this terrible accident that has affected the workers of our region, and we are confident of the success of this effort, especially as it concerns the lives and safety of the trapped workers." The communiqué also asked that people not travel to the area, so as to avoid hindering the efforts of the professionals now working at the San José mine.

With no news or confirmation of anything yet, the families were, of course, still worried. Some sat on the huge rocks that littered the site, others in chairs beneath sunshades they had brought for the wait. They gathered around Javier Castillo, a member of the miners' union. "We've been saying since 2003 that you can't work in this place," he said. "And yet, Sernageomin [Servicio Nacional de Geología y Minería, the government agency charged with overseeing working conditions in the mines] hasn't been able to shut it down." He clenched his fists in anger.

San Esteban Mining, the owner of the San José site, had faced repeated accusations of unsafe conditions for its workers. In 2007, the miners, along with labor unions from other mining sites that provided services to the company, lodged a lawsuit before the Court of Appeals against Sernageomin over the deaths of three miners in the San José and San Antonio sites. At the time, the workers asked that the mine be closed, but it didn't happen. Gino Cortez had been heading for lunch when a rockfall let loose on top of him. The forty-year-old man had worked as a fortifier, building shorings—the very work that sought to prevent the sort of cave-in that crippled him.

A FRESH ATTEMPT

But it was the last thing Javier Castillo said that made the biggest impression on the families: "There is no way out," he announced. Hands flew to open mouths, and people gaped in bewilderment. Others lowered their gaze to keep from showing their emotions.

Despite Castillo's dire revelation, rescuer Pedro Rivero wasn't giving up. He believed he could find a way out. He made ready for a new offensive in another sector of the mine. He would not accept defeat, nor would any other man on his team.

He geared up for another descent, this time with greater delicacy, for this other section of the shaft was even more fissured: ahead, behind, on all sides.

He decided to set up a base at the 967-meter level of the shaft and build a platform there to position the rescue equipment. From that point, they would mount a new attack on the lower levels in search of the thirty-three miners.

In a few hours, the wooden structure was constructed and properly anchored in the shaft. This base would support the weight of the rescuers, equipment, and tools.

Pedro was nervous and uneasy, concerned about the fate of the workers, whose whereabouts were still unknown. He knew he couldn't slacken his efforts in this rescue operation, perhaps the most important of his life.

One of his teammates, Pablo Ramírez, prepared to continue the descent. Side by side, they lowered down 656 feet of rope. The line stretched out of sight, lost in the void below. Then Ramírez stopped and called out a warning: "Pedro, the right side of the shaft is fractured. The duct is covered with debris, and farther down it may be worse."

Pedro didn't reply. They looked at each other in silence, sealing an unspoken pact between them to spare no effort and avoid no risk to find their fellow miners alive.

FAITH CRUMBLES

"It hurts me; it touches my heart," Laurence Golborne said to the family members, who had been informed during the night that there would probably be new developments within eight hours.

Minister Golborne, who found himself facing one of the most complex situations he could imagine, fought to stay calm. He zipped up his jacket and stretched his legs, which were tired after the plane trip from Ecuador and the hours standing vigil outside the mine.

"The news isn't good," he said, clearly upset. "We have not been able to make contact. In the shaft where rescuers were being lowered, further rockfalls occurred, and they had to get out to save their own lives."

The mood in the camp was grim. There were no more hopeful smiles. It was the first time that the family members started to understand just how complex the rescue operation would be and began to doubt that it was possible to reach the miners. Their faith seemed to ebb like the dust eddies in a midafternoon desert gust.

The operation was growing larger and looked as if it would take longer. A medevac mission was set up to carry any rescued miners to the hospital in Copiapó—and hope seemed to return slowly. A private company provided two helicopters and a Twin Otter airplane, fully outfitted with first-responder equipment.

At the site, other rescue workers, including the miner José Vega, worked in the duct, putting up wood shoring to stabilize the weakest areas. Undersecretary Pablo Wagner affirmed that air was circulating in the parts of the mine they could access, but no one knew whether the same was true farther below.

August 8

Expectations were so high the first two days that by day three many in the camp began to despair. Without anyone's noticing, much time had slipped away since the tunnel collapse. A figure dressed in black, with a large wooden cross at his throat, caught the attention of the family members. It was Gaspar Quintana Jorquera, the bishop of Copiapó.

"I ask you to put the lives of these men in the hands of our Lord, so that he may give them strength and hope," said the bishop. He comforted the families, who had already endured seventy-two hours of uncertainty and who now numbered more than two hundred people. "I know that you are living through hours of unimaginable anguish," said the cleric.

While the bishop addressed the families of the workers, the unexpected happened. San Esteban Mining, the company that owned the site, spoke about the accident for the first time. It was a belated attempt to confront the storm of criticism leveled at it by Chilean society for its negligence: for the lack of safety in the mine, of course, but even more for its incomprehensible callousness toward the family members when the events came to light.

"The event was unforeseeable," insisted Pedro Simunovic, the company's general manager. "It isn't that there was a delay in informing anyone. When the event began, we had to enter the mine to make certain of what was happening. Getting to the mine took us several hours," he emphasized.

Regarding the previous accidents, he said they had been resolved and that for this reason the company had been allowed to continue operating. He went on explaining what had become, for everyone, inexplicable.

THE OWNERS OF THE SAN JOSÉ MINE

The history of the San Esteban Mining Company, which owns the San José mine, is strewn with negligence, irregularities, and deadly tragedies. The company was founded in 1957 by a Hungarian immigrant, George Kemeny Letay. Upon his death, control of the company passed to his son Marcelo Kemeny and to Alejandro Bohn, who is Kemeny's brother-in-law. Both men are now under serious scrutiny for their role in creating the conditions that led to the disaster.

In 2007, the San José mine was closed by the government after a series of accidents that seriously injured some workers and killed three miners. In 2004, unstable conditions within the mine resulted in rocks breaking loose and falling down a mine shaft, burying a miner below. This notorious incident foretold future disaster, but the mine continued its operations uninterrupted.

Between 2007 and 2009, the San José mine was officially closed down by the Chilean government. But the company continued mining operations anyway, using subcontracted workers, and in July 2009 it was officially reopened by a motion of the Ministry of Health, even though that government body was unaware that the mining company had not followed regulations and put an emergency management system in place, and had not installed the recommended communications equipment. The state regulatory agency, Sernageomin, had only eighteen mine inspectors to enforce regulations throughout the entire industry, according to Senator Baldo Prokurica, a member of the Senate mining commitee. This number was apparently woefully inadequate for effectively overseeing the hundreds of mines throughout the country.

A HAIL OF ROCK AND DEBRIS

Many hours had passed since Pedro Rivero and Pablo Ramírez made their silent pact of unity. They descended into the mine again, three separate times, but weren't able to get through. Ramírez was ready to try descending through a ventilation duct. He lowered himself into the darkness, feeling confident and with the clear expectation of finding a way to reach the trapped miners farther down. After descending several feet, he paused to adjust the rope that held him suspended from above, then went down several feet more to a point where he was forced to stop.

A cascade of rock and gravel clattered down on Ramírez's helmet. The hail of mining debris came without any warning, and then the real cave-in started, to one side of the platform serving as a base for the rest of the rescue team. The violent impact mangled one of the communication cables. No one was injured, but it sent panic through the crew, while Ramírez still hung from a rope below them.

"Get me out of here, dumb-ass!" he shrieked up from the abyss. "The shit's flying all to hell!" Pedro Rivero could barely hear him, but he understood that he'd better order the hoist at once.

"Code red," Rivero's deep voice boomed.

Instantly, everyone began pulling hard and fast, desperate to save Ramírez before he was crushed by the downfall. They hauled and hauled on the rope until they were able to lift him aboveground. Finally he stood, as though frozen, on the platform.

The standard procedure in a rockfall is to get to the surface as fast as possible, because at any moment, a new cave-in can take out the entire group.

Some crew members grabbed their gear before leaving, but most of them retreated at a dead run, leaving tools and ropes scattered across the platform.

And with that, unhappily, the latest attempt to reach the trapped miners through the ventilation shaft had failed.

"WE'RE NOT GOING TO LEAVE
THEM ALL ALONE"

Unlike the private executives who refused to give a press confer-
ence, Minister Laurence Golborne faced the journalists with a
steady gaze. "We're talking about several days, probably more than
a week," he acknowledged, referring to a possible delay in locating
the miners.

Still, things could always get worse. And this new collapse, bigger than all the others before it, ended up burying any possiblity of
using the ventilation shaft.

The echo of the rockfall and the sudden scramble of rescuers
fleeing the mine demolished what remained of the families' hopes.
With the tube blocked, they had to start again from square one,
with a whole new plan. Things looked grim.

They decided that six rigs, four of which were already on site,
would drill from different points on the San José site, at rates rang-
ing from 164 to 328 feet per day, depending on the type of machine
and the hardness of the rock encountered. It was a matter of drilling
multiple test holes, the object being to reach the workers, whose
status was still unknown, and supply them with oxygen, water, and
food before it was too late.

They chose to drill several boreholes simultaneously in order
to minimize the margin of error. The efforts never slowed for even
a second. Though despair ate at the souls of the waiting families,
those responsible for the rescue task soldiered on. The enormous
trucks pulling the drill rigs flung gravel up in the air as they made
their way onto the site. The families cheered, crowding around to
see up close the technology on which they now pinned their hopes.

Soon the noise of the drilling engines filled the site.

"We're not going to leave them all alone," President Sebastián Piñera announced with confidence. Having broken off his trip to Ecuador, he was now at the disaster zone. "In my meeting with the families, I reaffirmed to them our absolute commitment to do everything humanly possible and spare no effort, no rescource, to bring back alive our thirty-three fellow countrymen trapped in this mine." Privately he acknowledged to his advisers that "the situation was difficult." In his mind, one always expected the worst while striving for the best.

Meanwhile, some of the miners' families protested at not being allowed to participate in the meeting with the president, which took place at night in the offices the company kept at the San José mine and to which only a small group had access.

"We should *all* be in these meetings—not just a small group," complained Darwin Cortez, whose brother, Pedro, had disappeared in the mine. "Here everybody is desperate to get news of our people," he insisted.

Faced with these objections, President Piñera explained, "It was better to meet with the representatives of the family members, to have a closer, more in-depth conversation, to analyze all the options and to convey to them the government's commitment. And I hope that the families will be able to speak, and are doing so today, with the others, so that all can be better informed."

August 9

Four days had passed since the collapse. As time seemed to stagnate, Chile's undersecretary of foreign relations, Fernando

Schmidt, held an important meeting at the foreign ministry in Santiago. He had invited the representatives of other countries with with long histories of mining technology. The chiefs of the delegations of Australia, Canada, the United States, and South Africa arrived punctually.

Briefly, the undersecretary detailed the situation in the San José mine. He said, "If there is any technology, any knowledge, that can help us, we are asking you for it." The diplomats left the building on Teatinos Street in the center of the capital, intent on consulting the experts of their respective countries, including the big engineering and tunneling companies.

That same day, President Piñera reiterated that "everything humanly possible" would be done to bring the miners out alive. He asked Codelco, the state mining concern, to recruit its best people. The largest subterranean mine in the world was the El Teniente mine, located in Rancagua, Chile. The head engineer there, André Sougarret, answered the call.

From that moment on, under presidential mandate, this civil engineer would carry on his shoulders the job of returning the Atacama thirty-three to the surface.

Those close to him described him as a professional executive, well liked and possessing great charisma, who excelled in whatever task he took on.

He had been chosen for this herculean job because he was able to build excellent working teams and because he knew how to listen to his subordinates—a skill that inspired confidence and allowed him to work very closely with his people.

Sougarret, the fifth of six brothers, was married with three daughters. He had twenty-three years' professional experience, during which he had distinguished himself in leading deep excavations. Moreover, he had participated in the rescue efforts after three collapses at El Teniente.

PRECARIOUS SAFETY CONDITIONS

Ironically, the national holiday celebrating the mining profession fell on day five of the search. On the eve of August 10, the Day of the Miner, representatives of San Esteban Mining reported that one of the drilling rigs had reached a depth of 656 feet inside the mine—almost double the estimated depth for a day's drilling. The objective was to install tubes through which to lower provisions, for at this point, with the passing days, the trapped men's rations were likely at an end.

Specifically, the latest efforts were focused on drilling small holes to deliver food to the miners—in the event they were found alive—while working out an alternative way to free them from the subterranean deposit of copper and gold in the collapsed shaft.

It was inevitable that questions on mining safety would come to the fore. The tragedy, which kept Chile and all the world in suspense, brought to light the precarious safety conditions of the small mines operating in the north of the country, where hundreds of miners risked their lives daily.

"We have been able to verify, with the case of the San José mine, that in many mineral extraction operations the safety of our workers is not getting the consideration it should," Piñera himself ackowledged. He promised to quintuple the budget of Sernageomin, with the aim of augmenting the number of inspectors.

GRAY RAIN

Despite this good news, the Day of the Miner turned miserable for the families. In the Chilean desert, rain is uncommon, but when there is enough precipitation, it comes down with a fury. That day the sky was leaden, turning into an ominous premonition for the families. With a loud racket, the wind whipped the tents that had been pitched, one by one, within a few yards of the mine, and the gusts mixed water and dirt into mud. The storm sprang up suddenly and turned the newly established camp into a quagmire. Everyone got down to the work of saving their little spaces from the flooding.

Officials of the National Emergency Office (ONEMI), walking over the boards that people had laid out in a makeshift path over the water and mud, handed out mattresses, blankets, camp stoves, and gas cylinders that they brought for the families that stoically remained on the site. They also brought big rolls of polyethylene for lining the tents.

The national director of ONEMI, Vicente Núñez, had come to assess the needs on the ground in accordance with the minister of the interior's directive for attention to the people affected by the emergency. "We are prepared to lend the full support of ONEMI in the tasks of operational coordination and rescue," Núñez explained. "Our main concern is the miners and the protection and shelter of their families, as well as managing the needed technical elements and materials."

A few yards away, Luis Briceño, police chief of Atacama's Third Region, sat down on a plastic chair, contemplating the mobile precinct station that would remain at the mine for the duration of the rescue effort. This facility would be manned twenty-four hours a day by a police captain tasked with attending to the needs of the miners' families. This included providing them with a way to stay in touch with family and friends, both locally and outside the region, via six cell phones that the police had made available.

Also, infantry soldiers, both in jeeps and on horseback, were on site to keep order and security among the more than five hundred people who had come there to await the rescue of their loved ones.

One of the specters haunting the camp was money. The families had no way to support themselves at the mine, and so they turned to San Esteban. They asked the management to give them the wages of their trapped loved ones. In this campaign they had the support of Socialist Party senator Isabel Allende, daughter of the assassinated president, Salvador Allende.

"Concerning the miners' wages, the San Esteban company should continue paying each of its workers in full, until such time as Sernageomin decrees that the mine should be closed," the senator argued. "Behind each trapped worker is a reality—there are debts and commitments that they must honor."

After a whole week, the paychecks never came, and the people's sense of outrage grew.

The discontent in the camp also grew over the political debates that the financial situation generated. The families felt that time and resources were being wasted on useless wrangling while their loved ones remained trapped underground. And the men, they emphatically reminded one and all, were the only thing that really mattered.

In this situation, Representative Giovanni Calderón, of the Independent Democratic Union, the right-wing party that supported the current president, said, "Since the reopening of the mine, in 2007, there have been more fatal accidents. On this basis, we are going to request an investigation into why Sernageomin allowed the mine to reopen. This is an investigation that should be conducted before the comptroller general's office, to determine whether the authorities at the time may have committed a crime by ignoring some legal regulation."

At the same time, Chile's attorney general, Sabas Chahuán, reported the opening of an investigation into the accident. The district attorney of the city of Caldera, near Copiapó, was placed in charge. Caldera was where the investigators gathered to determine whether the accident at the San José site constituted a crime of involuntary manslaughter or injury.

Regarding the legal representatives of the mining company, Chahuán was prepared to investigate any negligence that may have occurred in the chain of command required for this type of mine. "The only way to know whether a crime happened is to investigate," he said.

Regarding the later collapse that occurred during the early rescue efforts, the attorney general did not rule out the possibility of a proceeding to determine whether some criminal act might also be attributable to the owners of the San Esteban company.

It was a day of much activity on several fronts: stepped-up rescue efforts by Codelco personnel; President Sebastián Piñera's decision to ask for the resignation of three Sernageomin officials, including National Director Alejandro Vio, over anomalies in the certification of the San José mine; and Minister of Mining Laurence Golborne's official request, before the comptroller general's office, for an internal indictment within Sernageomin.

The investigation would seek to "determine if this has indeed been in accordance with the existing rules and within the powers that the service [Sernageomin] has or should have carried out," Golborne affirmed.

Golborne declared that he would await the report from the comptroller general's office, and that if any sort of criminal act should come to light, "let there be no doubt that we will act with all the force that the government possesses."

A CONTROVERSIAL REOPENING

Although earlier in the San José mine's century-long history it produced silver, these days it extracted mainly gold and copper.

In the 1980s, the mine passed to the hands of the San Esteban Primera Corporation, founded by the Hungarian immigrant George Kemeny Letay. Kemeny died in 2000, and since then the administration fell to his sons, Marcelo (an architect) and Emérico (a photographer). Emérico died in 2005 from lung cancer.

Although the sign at the mine's entrance, not so far from where the thirty-three were trapped, proclaimed, "TOGETHER, WE WORK SAFELY," the San José mine had, in fact, seen several accidents since the year 2000, finally forcing the government to shut it down. It stayed closed from March 2007 to July 2009, when Sernageomin authorized its reopening.

Since 2003, the employees had criticized the dangerous working conditions in the San José site, where a new production method imposed by the company had weakened the walls, eventually triggering the catastrophic collapse.

CRISTINA'S PHOTOS

And through it all, the numbers of people standing vigil at the impromptu settlement bordering the San José mine continued to grow.

Cristina Núñez, a petite woman who always dressed in a track suit and wore a jockey cap to protect her face from the sun, walked with hurried steps between the tents. Recently the inhabitants had nicknamed the settlement Camp Hope, to affirm their unshakable faith, and one of the people to suggest the name was none other than this woman in the jockey cap.

Cristina wanted to convene all the family members to agree on a common stance on all the proposals by the authorities. Officials wanted the people to vacate the site, saying it was for their own security, but she and many others suspected that it would be better to stay put. They wanted to see for themselves the progress of the search.

"Distrust isn't always a bad thing. We have the right—those are our men," she said in support of the decision to stay, a position that met with general approval from the others.

The small woman with the trademark long black mane hanging down her back also had erected a little altar on a rock, in homage to her Claudio. Claudio Yáñez, an explosives operator, had been working at the mine for eight months. His dark, thin face appeared in a photo that Cristina had placed at the top of the shrine, along with other personal effects.

These little shrines, made in memory of those who had died on the job, whether in a mine or on the highway, were commonplace in northern Chile. They were known as "*animitas,*" and those of truck drivers also often included vehicle license plates in the decor. Cristina certainly hoped for a better fate for her husband, although she remained deeply concerned.

"He told me it was unsafe," Cristina recalled. "One time a fifty-kilo rock fell right beside him. If he hadn't jumped, it would have killed him."

She believed that her husband had been born under a lucky star and that somehow, although there were no guarantees, he was alive. She dreamed of his escape and of returning to their house in San Bernardo, a densely populated city in the Santiago metro area, where he might find a less dangerous job.

Picking her way among the rocks, Cristina kept her hand in the right pocket of her jacket, where she clutched her most cherished possession: a set of four photos of Claudio's face. When the mine required him to get photographed for his accreditation as a worker, he had gotten ten images for a volume discount. He gave six to his employers; the rest were a present for his wife.

"No, I don't look at them much, because it makes me sad," she confessed, "but at night I gaze at them."

CHAPTER THREE

MEN CRY, TOO

During their meeting at the camp, the families put up a little figurine of St. Lawrence, patron saint of miners, on one of the hillocks that surrounded the tents. From now on, this would be the place where they gathered to pray for their loved ones.

The spot bore witness to a surprising scene. In the middle of a commemoration mass, the police captain, Rodrigo Berger, burst into tears. Seeing this tough cop in uniform crying, the family members instinctively drew around him in support. He was no longer the authority in charge; he was another human being, like everyone else.

"I couldn't bear it," the captain explained later. "Nothing like that had ever happened to me." He said he was overcome with emotion on seeing a child cry in its mother's arms when its father's name was mentioned in the sermon. That spontaneous reaction broke the policeman's heart. "It made me sad, waiting together with these people for five days, living side by side with them through this rescue."

From that moment on, Berger's gift of humanity made him a favorite of the family members. Now, whenever a question requiring consultation came up, they asked to speak with Captain Berger. To them, he felt like another member of the camp, one of their own.

A COURAGEOUS WOMAN

Day 9

The rescue efforts seemed to be moving ahead, but still there was no sign of the trapped men. Life at Camp Hope took on a more defined form, though it was still full of surprises. That was how things stood when Carola Narváez arrived at the site.

The woman came from Talcahuano, in southern Chile, one of the areas hardest hit by the earthquake and resulting tsunami that had devastated much of the region in February 2010.

And that was precisely the reason, Carola explained, that her husband, Raúl Bustos, was inside the mine on that fateful Thursday, August 5.

Her husband had worked as a mechanic in the shipyards of Asmar (the naval arsenal), but since the earthquake and tsunami in February he had been without a job. He had to find something, and that was how he wound up at the mine. Raúl hadn't wanted to leave home, but he had no other option.

The journey from Carola's home to the mine took a whole day, and when she got there she slept in her car. She said it didn't matter, that she did it all in her eagerness to find Raúl alive—even if there was only a tiny chance of his being saved.

"It's not even his job to be inside the mine," she said. "I told him to stay out of there, but to do the best job, he said he sometimes needed to go down and inspect the vehicles in the tunnel."

The new arrival got on well with Mario Gómez's family, the largest group in the settlement. Camped in a broad area beside the tent used by the press and near the lunchroom, the Gómez-Ramírez family numbered forty-six persons on the registry of the territorial administration, which kept strict control over who entered the site.

"We're all relatives—it's just that my mom and dad each have eight brothers and sisters, so we're all here together: children, brothers and sisters, nieces and nephews," said Roxana, one of the ore truck driver's children. The Gómezes had brought their small children, who romped among the rocks and played tirelessly at marbles.

Meanwhile, far from these scenes, Sernageomin fired its director, Alejandro Vio. Perhaps out of solidarity with the miners or to preserve his own reputation, Vio showed up at the camp. After a half-hour meeting with the minister of mining, the ex-director, responding to criticism of him and the agency, read a communiqué in which he emphasized that "the most important thing is the rescue of the thirty-three workers. . . . Although I am a technical authority, I have been asked to bear political responsibility. I do have a clear conscience . . . I will continue to cooperate in any way I can."

Socialist senator Su Camilo Escalona criticized the firing, as did other state officials: "It is easy to lay responsibility on civil servants of minor rank, who are always criticized as being part of an inefficient state bureaucracy," he argued. "And yet, they must rely on minimal facilities to accomplish their tasks—and they don't get paid for working overtime."

President Piñera's government had a ready rebuttal to objections over Vio's firing: They pointed to his negligence in granting permission for the San José mine, his lack of control over prior situations, and his nonexistent oversight of the work in the mine. As director of Sernageomin, he was the administrative official responsible in the disaster.

Meanwhile, the families at Camp Hope saw the political debates and mutual blame games as pointless. They only wanted everyone to stop the accusations and focus on getting their men out of the mine alive.

"THE NIÑOS ARE STRONG"

Cristian Ulloga grabbed his hard hat, glanced at his watch, and looked at the monstrous rock blocking the mine entrance. Now it was time to use the drill to bore some holes and place the explosive charges.

Like dozens of other small mines in the Atacama region, the San Javier mine, in Tierra Amarilla, just over nine miles south of Copiapó, kept right on working after the collapse at the San José site.

Cristian continued working with all his energies—or most of them, anyway, since part of him was somewhere else: with his lost comrades.

"I think they're alive," he said, sure of his words. "There's a good chance that they are, and in a while we will see them again. They're miners; they know what to do. What's more, they have some seasoned guys down there to create order, and the younger ones have the energy to pull them all through," he added, with obvious pride in his profession.

Still, days after the San José mine shaft had fallen in without any notice, hope was beginning to flag in many. But then there was Cristian, who, at twenty-six, had a personal story that drove him to believe that his colleagues would survive.

He had come close to being one of them. Carlos Barrios, his friend since boyhood, had offered him a job at the San José mine. Now Carlos was one of the thirty-three who were trapped.

"The Sunday before the collapse, I was ready to go," Cristian said. "It made sense. The pay was 700,000 pesos [$1,400]—better than anything else in the area—because everybody knew that hill wasn't safe, so that was what they had to do to get workers."

Cristian said he had thought hard about it. But his father, Luis Ulloga, an experienced miner, had told him to turn down the attractive offer.

Luis Ulloga joined the conversation. He wanted to express his concern for the trapped workers.

"Everyone knows that a lot of accidents have happened in the San José—it was dangerous," declared Luis. His skin was dark from his work in the mine, which he reached by a narrow dirt road skirting a steep hillside.

When a truck went up, they had to give a signal to keep other vehicles off the road, which could handle traffic only one way at a time. At the San Javier site, everyone went up or down the road at the same time. Such was the risk; such was the mining routine.

Cristian Ulloga learned of the collapse the morning after it happened, when his wife turned on the television. He thought at once of Carlos, his friend since they were twelve.

"The *niños* are strong," he repeated with conviction.

Another worker from the San Javier site appeared: Pedro Cifuentes, who ran one of the machines there. He came over, interested in giving his version.

The group listened eagerly. Several set aside their normal tasks, and the conversation grew.

Pedro looked at the others and spoke fast before anyone could interrupt him. He said he knew that another of the trapped men was a well-known former professional soccer player, Franklin Lobos, because, besides working in the mine, Lobos drove a taxi in the city.

Like Cristian, Cifuentes believed that the thirty-three were still alive.

"The boys may be banged up, but they're alive," he said stubbornly.

It's the same stubborn determination that one could see on any street corner in Copiapó, among those who knew the trapped men and those who had never exchanged so much as a word with them but who were nonetheless steeped in the feeling of solidarity that permeated the city.

That stubbornness, they say, is a fundamental part of Chilean mining culture. They are stubborn, they admit it, and they like being that way—they feel respected for it.

This social trait comes through in any conversation around the usual bartop or table where they get together for beer, after payday or *el suple,* as they call the advance on their wages that usually comes in the middle of the month. Or they may socialize at any soda fountain or *schopería,* as they call bars these days, where they fill the place, asking right away for a table all their own.

At those times, when nothing is left in the beer bottles but a little fluff of foam, everyone knows best about whatever the topic is—

typically soccer, women, or work. If there are ten workers, chances are there'll be ten conflicting opinions.

At home, too, this same stubborn determination reigns. The miner always "knows what he wants." As the head of the house, he speaks and the others listen. It's been that way for generations and generations. A miner might need to argue different views with his fellow miners, but his family always takes him seriously. They know the miner as trustworthy, faithful, and honest. The truth, whether real or imagined, lies in the deeds. A miner's stubbornness also has its limits. It softens, quite appropriately, inside the mines. Since the work is so delicate and any mistake could cause a fatal accident, everyone, every *viejo,* bows to the knowledge and experience of the other.

"*Viejo*" is the word they all use for each other in the Chilean mines, especially in the north. It is a general nickname, but not just anybody is a *viejo*—first you have to know something about mining.

Now the *viejos* trapped in the San José mine were called "*niños,*" as a way of expressing the vulnerability and fragility in which they found themselves.

FAITH IN A MIRACLE

In Camp Hope they didn't believe in the dire predictions, or that time was against the miners, or that the vultures that wheeled in great circles above the mine meant trouble, although some attributed the presence of these birds to an ability to smell misfortune from miles away.

Neither the presence of the vultures nor the difficulties of the situation dimmed their hopes.

While the rigs kept drilling away at the San José site, far from the streets and buildings of downtown Copiapó, in the hills and mines where darkness reigned, the mining families believed that a miracle would occur.

"NOW THIS IS MY HOME"

Night is a special time in the desert. At midafternoon the scorching sun, which has burned all day, begins to wane and is replaced by a cold that seeps right through your bones. Then comes the gray mantle of the *camanchaca,* that damp sea mist that descends at twilight and shortens visibility to only a few feet ahead. After doing their own thing during the day, the families would meet up and share their joys and recent misfortunes.

Once the darkness and cold set in, a quick trip among the tents found most of them empty, with hardly anyone sitting by the firesides.

The people chose to meet in the lunchroom, the biggest of the structures, where there were tables, an image of St. Lawrence, and a television (permanent company in the solitude of Atacama).

There they could eat and watch the news, which invariably began each evening's broadcast with an update of the mine. Many of those present gazed, riveted, at the screen.

Others preferred to be in front of the image or the banner that reminded them of the loved one they longed to see. In front of Yonni Barrios's photo stood his brother and Marta Salinas, his wife. They prayed in front of his photograph, the vapor of their breath visible in the cold and clouding their faces. In a nightly custom, they lit candles, leaving them in the rocks to illuminate

Yonni's face, which gazed back at them in silence with a questioning look.

Not all was harmonious. Evening had fallen, and shouting could be heard in the distance.

A woman of the Segovia family was yelling at one of the occupants of the tent of Pedro Cortez's family, pitched just opposite. The discussion concerned a man who apparently looked a bit too often at his neighbor as she passed by.

After that incident, the two families quit speaking.

Meanwhile, amid the slow days of waiting, the hours were broken up by the occasional visitors who dared make the trip up this inhospitable hill, to where the camp lived and breathed.

Over the tents, on the hillock at the mine's entrance, a sign, which now appeared strangely mocking and cynical, welcomed the workers of San Esteban Mining: TOGETHER, WE WILL MAKE OUR WORK SAFE.

This was not the only sign to grab the visitors' attention. For example, the students of a Copiapó high school, dressed in their severe uniforms, paraded through the only street in the camp, amid the cheers and thanks of the families, with a canvas bearing the poetic message: "OUR MINERS WILL RETURN AS HEROES FROM THE DEPTHS."

On weekends, it was obligatory to listen to the powerful loudspeakers of the evangelists, who, making the most of their majority among the families, set up on an improvised plywood platform, converted into a stage for all the activities at the site.

Their competition was an ancient woman with white hair, who always dressed in black and who, Bible in hand, walked back and forth through the middle of Camp Hope.

"Jehova sent this punishment," she repeated endlessly, interspersing this with selected readings from the Bible. After the initial curiosity wore off, few people attended her jeremiahs, but this did not sway her from coming every Sunday to the mine, to perform the religious rite that nourished faith.

CHAPTER FOUR

ANGUISH VERSUS LAUGHTER

An unexpected character to join the vigil was Rolly the Clown, who came from Calama, the place most emblematic of Chilean mining. Calama was the city of the workers in the Chuquicamata mine, which for a few years has been the biggest open-pit copper mine in the world.

Rolly's personal challenge was to raise the spirits of the children in the camp. He had come for only a few days, but at the children's request, he was ready to make the camp his home.

This traveling artist could well be a symbol of Camp Hope. With his sky-blue jacket, red t-shirt, and striped pants, he went everywhere, with no permission asked or needed. He shared chocolates and candies that he got as donations. Everyone knew him. The foreign media ran articles about the "clown of hope."

Reflecting on the media whirlwind that had taken over the site, Rolly, sitting across from a European journalist, said gravely, "At some point, this thing got out of hand."

Rolly's words, even said with his mask in place, revealed the serious side beyond the jocular face he put on to make the public laugh. His name was Rolando González, and he worked in Calama for a company that provided workers for the El Abra mine. In his free time he dressed as a clown.

He was constantly traveling to places where people needed him, where tears needed turning first to a wisp of a smile and then into loud, healing guffaws. And Rolando always did it for free.

That is to say, it was his calling, his passion, the essential aspect of his life: He rushed to scenes of disaster, entertaining little ones after the Tocopilla earthquake and again after the quake of February 27, 2010, catalogued as one of the greatest ever registered in the earth's history.

HOPE TAKES SHAPE

As the days passed at Camp Hope, the organizing efforts were bearing fruit.

The township of Tierra Amarilla used a shipping container and an awning as a day care and school for the children who were staying on the site with their families. The makeshift school had fifteen children, aged one to twelve years old.

"It helps distract the children a little from the tense atmosphere that surrounds them," explained Ana Funes, the social worker in charge of the facility.

There, on the settled dust and by the will of Ana, the children drew while the trucks of the drill rig operators drove ceaselessly to and from the mine.

Katherine was a good example of the children at the site: a sweet kid with long black locks, her face beautifully etched with the indigenous attributes of tribes from the South American highlands.

Those strong features, inherited mainly from the Aymara and Inca ethnic groups, had changed little despite later racial admixtures: oval face, prominent cheekbones, thick black hair, glowing dark skin.

The works of art by the children of Camp Hope had a recurring theme. Katherine, for example, drew an immense face that covered the entire page, with two huge tears. Others drew the flags on the hill, single points of color amid the desert gray landscape that surrounded them.

Constanza, five years old, sketched a bearded figure dressed in a white tunic. "I, God, am with you all," she wrote in wobbly letters with an unsteady hand.

Bastián, grandson of Mario Gómez, made the most striking drawing: a happy, smiling face bordered by a rainbow extending between the mountains.

Asked, "And what's that?" he replied, "That's how I'm going to be when they get my grandpa out."

GAMES, TRADITIONAL AND MODERN

Evidently, the children didn't spend all their time drawing. They also played games, interweaving traditions of an earlier time with the technological devices that their parents had bought in local markets.

In this innocent play, portable PlayStations coexisted in apparent harmony with local traditions such as the practice of *payaya*, a children's game that consisted of holding five small pebbles in one's palm, tossing them up in the air, and catching them on the back of the hand. The winner was the one who caught the most pebbles without dropping any.

They also amused themselves playing *luche,* an Atacaman version of hopscotch, in which the children throw a rock at the design of a cross formed of numbered squares. They hop on one foot, advancing square by square, trying to keep their balance from one end to the other, and then go back the same way.

There was also a sort of jump rope, a passion of the little girls. A rubber band was stretched a foot and a half above the ground, and those holding it made figures of increasing complexity, widening or narrowing it at whim. Then, combining physical dexterity, rhythm, and speed, one had to jump over it again and again.

MUTINY IN THE CAMP

Every evening the rescuers would come back to the camp after a long day of drilling. The families ran out to greet them and, by their tired faces, would know right away that there were no developments to report. The lack of progress was taking its toll on the families. While the children amused themselves, the adults argued.

"More than an accident—it was a crime," declared Angélica Álvarez.

Her husband, Edison Peña, was trapped in the mine. She walked around Camp Hope to pass the time. She didn't like to talk with the news people, but sometimes she needed to vent.

"The owners of the mine are to blame. They had been told it wasn't safe, but they didn't pay any mind," she said angrily. In a calmer moment, she reflected, "Now is the time to pray; the rest can wait till later."

Angélica referred to the investigation being conducted by the Caldera district attorney and the Atacama regional public ministry into working conditions at the San José mine.

These public figures came to the site several times during the drilling effort to talk with the rig operators and the family members. Their work began with a statement by Gino Cortez, a miner who had lost a leg in a work accident on July 11, almost a month before the collapse. Gino's case became the tip of the thread that the prosecutors followed to investigate San Esteban Mining.

No one knew which turn the investigation would take. If the workers emerged alive, it might turn into nothing more than a complaint. If their beloved never returned from the depths and the families received death benefits, San Estaban Mining would probably face criminal charges.

Gino himself appeared from time to time at the camp, propped on his crutches, and made it his task to banish the darkest thoughts.

"Don't get discouraged," he said. "The *niños* are fine. They should be feeling the drilling now, and that will perk their spirits up. I remember when I had my accident how fast everyone mobilized to help me. That's how we miners are: always up for a joke, but when somebody needs help, we react in a second."

The days went by, and although there were still no signs of life from inside the mine, the rescuers never paused. They searched intently in different parts of the mine, drilling twenty-four hours a day in three eight-hour shifts.

The leaders of the Chilean Safety Association were sure that the drills would reach the miners sooner rather than later, and based on that supposition, they already had their first message ready to send to the trapped men.

The first line read, "We are with you."

The message was really a set of instructions that would be sent down along with three *palomas* (plastic capsules): one full of water bottles, one with medication, and one with sugar water:

1. —*Take one bottle of water slowly, in small sips so as not to vomit. Take a second bottle about fifteen minutes later. Continue with a bottle every fifteen minutes.*
2. —*With the water of the second bottle, take a tablet of omeprazol (medication to cover the walls of the stomach to prevent overacidity and gastritis), which will come in an additional bottle.*
3. —*We will send you more bottles of water, which you should sip slowly.*
4. —*After that, we will send you a bottle of sugar water. Drink it slowly, in sips.*

Some family members were not so optimistic, however. In fact, some declared openly that they had little confidence in the work the authorities were doing. Time was slipping away, they said, with nothing to show for it.

The family of Florencio and Renán Ávalos even sought help from Juan Ramírez, an experienced rescuer who had worked in several Atacaman mines extracting copper, gold, and silver using traditional methods.

The man turned up at the camp and met with all the families. He assured them that he and four colleagues were ready to enter the mine through the principal tunnel, down to the huge rockfall of 700,000 tons obstructing the escape.

He insisted that with 660 pounds of basic equipment, including picks, jackhammers, and explosives, and with a week's work, his team could punch a hole through the damned mess and get to the miners.

"We're used to working in terrible conditions and risking our lives," he said to the astonished families, who latched on at once to his plan. They presented the idea to those in charge of the rescue effort at the daily meeting held at six each evening.

After hearing them, the undersecretary of mining, Felipe Infante, rejected the risky plan right away. They were not about to have the little group go in and work in that unstable mine, which could give way and swallow up thirty-eight bodies in a single gulp. For now they were ruling out the rescuers' plan.

Several family members walked indignantly out of the meeting.

"They don't want to hear us," they complained. "They just blindly trust in their machines and don't see time ticking away."

Their anger smoldered and spread, and they all gathered at the entrance to the mine.

Cries of "We want the rescuers . . . we want the rescuers!" drowned out the noise of everything but the drilling rigs.

There were tears and recriminations. The scene resembled nothing so much as a mutiny.

"We're ready to go down, with or without authorization," Juan Ramírez proclaimed, provoking loud applause and drawing the police off their patrols around the perimeter of the mine. They began to keep a close eye on this unexpected civil demonstration.

The fervor with which Ramírez defended his ideas might sound like obsession or sheer pig-headedness, but behind it lay a certain rationality. He believed that the miners had already withstood so much adversity. Precious days were passing and little progress had been made. The men trapped 2,300 feet below the surface had risked everything, and to gain so little. And, sadly, this tragic payoff was part of the country's mining history. Yet, Chilean miners had managed, despite constant misfortune, to recover, and even to triumph, both above and below ground.

Meanwhile, the police managed to calm the protesters.

Cooler tempers prevailed the next day. Another meeting was held in which the authorities explained again that the Ramírez plan was little more than a hot-headed adventure that might put the five men in mortal danger and might further set back the chances of rescuing the miners. Meanwhile, the government-led effort was closing in on the safety area where the miners were probably waiting. The families needed to understand that they had to wait for the results of the drilling. Of course everyone appreciated the rescuers' willingness to help out.

This was the first crisis at Camp Hope and it had been settled smoothly. But, just in case, Juan Ramírez and his team stayed on, circling like birds over the camp, with their equipment at the ready.

And if the drills should fail, well . . .

INEQUITY, THE OTHER TRAGEDY

The accident involving the thirty-three Chilean miners laid bare the paradoxes of a country that distinguished itself as one of the most developed in all Latin America and at the same time denied—or perhaps did not want to see—the social inequalities so obvious in many parts of the country.

The massive operations that Chilean companies had built throughout the region, estimated at 50 billion dollars since 1990, coexisted side by side with millions of people who still subsisted on less than two dollars a day, according to the International Labor Organization.

The tragedy in the Atacama Desert sparked indignation and astonishment, but its most shocking effect was the revelation of how, before and during the disaster, the owners of the business and the national authorities barely considered the miners' safety. Indeed, San Esteban Mining, the owner of the San José copper and gold mine, wouldn't even pay the wages of its trapped employees or provide help for their families.

As the rescue operation was in full swing, the government, the courts of justice, and the National Congress were investigating, at every level, just how this company had managed to reopen and resume operations in a mine forced to close after repeated accidents caused by unsafe working conditions.

President Piñera, after acknowledging the problems confronting workers in the Chilean countryside as well as in construction and mining, gathered a team to study labor reforms.

The statistics were staggering. In the past five years, there had been twenty-three fatalities in mines throughout the country. Chile's miners worked an average of fifty-one hours a week—more than any other sector of the economy, according to official figures.

CONCENTRATION OF WEALTH

In Chile in 2010, an elite few controlled 47 percent of the assets of companies listed on the Santiago Stock Exchange. Well-known and recognized families such as Luksic, Angelini, Matte, and Solari ran the principal businesses. A small group of politicians also had amassed large fortunes, including the former Renovación Nacional senator Marcos Cariola, the Democratic-Christian senator and former president of the republic Eduardo Frei Ruíz Tagle, and the current president, Sebastián Piñera. Together these two groups represented a combined 9.16 percent of the GDP in 2004, and 12.49 percent of the GDP in 2008.

For a more complete financial picture, along with the 47 percent of assets of their various businesses, add the concentrated economic activities such as pension fund administration, private health firms, and the financial system.

International analysts still had an image of a successful Chile with the high growth rates of the 1990s, which had brought down poverty and improved the quality of life for most of the population. But for other economic observers, this wasn't enough. It was just as important to reduce the huge gulf of wealth between the classes by guaranteeing health and employment for all.

Since its return to democracy in 1990, Chile, as a full member of the Organisation for Economic Co-operation and Development (OECD)—the club of the industrialized nations—built a road to development paved with sustained growth, inflation control, and the elimination of public debt. The labor unions struggled to get a dignified wage, safe working conditions, and development of their trades. But with such an unequal concentration of wealth, and with an electoral system that denied proportional representation of political choices other than the two principal parties, the voices of the workers simply were not heard.

In the wake of the mine disaster, demonstrators took to the streets. There had been frequent protests in the capital and strikes by civil servants and other workers in the public sector—a group that had long since grown tired of broken campaign promises.

THE MINERS' COMPENSATION

The Chilean government refused to assume responsibility for paying the trapped miners' salaries, even while family members and union organizers urged the authorities to pay the men's wages as soon as possible.

Alejandro Bohn, one of the owners of the San José mine, admitted that the thirty-three miners were not covered by any type of insurance that would provide compensation for being trapped. The owners of the mine had provided no insurance for their workers, toiling at such a hazardous profession. In an attempt to rationalize this negligence, he said that the insurance coverage would have been "very low" anyway. He added that it would be "difficult" to pay the men their salaries because all of the mining operations at San José were shut down—such a "prolonged closing" of the mine led to "economic deterioration" for the company, from which they had not recovered, according to Bohn.

The families and labor leaders considered various options to try and get compensation for the miners, including an award for damages, but the authorities and the owners of the mine were "focused on rescuing the miners," according to Bohn. "It all depends on how our talks go with the government. Our company is very small, and our only mine in active operation was the one at San José," Bohn said from his comfortable office in downtown

Santiago. Even though the San Esteban Mining Company owned other mines, Bohn pointed out that getting them up and running would require an initial investment that at the time his company could not afford to make.

LABOR UNIONS FOR MINERS

The situation the thirty-three miners were in, completely unprotected by labor laws, was actually much worse than the reality for most of Chile's miners. In contrast, most miners at the major mines belong to strong labor unions, and the companies follow international safety regulations and sustainable business practices, according to official data from the Mining Council of Chile, the entity that ultimately oversees all matters related to the mining industry.

At most mines, the salaries are high, the benefits are good, and the miners are well-trained. They take courses in safety and are well-equipped with high-quality hard hats, gloves, and tools. They are even rewarded with bonus pay if their mines have perfect safety records.

In sheer numbers, the mining industry has fewer accidents than other industries, but the ones that do occur are more serious and are more often fatal. The labor conflicts the bigger mining companies tend to have are generally with subcontracted workers who are not directly employed by them; even if these subcontractors earn generous wages, after a while they demand to be put directly on the company's payroll so they can be eligible for benefits.

At the smaller and mid-size companies, the panorama is different. In general they don't invest in reinforcing the mines for safety, which became abundantly clear to the whole world with the collapse at the San José mine.

The National Service of Geology and Mining (Sernageomin) keeps detailed records on the mining industry. They carry out inspections to analyze health, safety, and overall labor conditions, although due to a small number of inspectors there is often a substantial backlog.

The National Copper Corporation (Codelco) produces a third of Chile's national copper output. The rest is under the control of private foreign companies. Codelco generates more income than any other company in the country. For that reason, many young people decide to work in the mines, and because the salaries are relatively good, few ever eventually leave that line of work. And some miners seem to have come to their profession by vocation. While they were waiting for the rescue to get underway, many family members of the trapped miners have said that in spite of the serious risks mining entails, they hadn't heard any of the thirty-three say they intended to give up mining and leave the desert lands of Atacama far behind.

But mining also comes with serious health risks. The most common illness that miners develop is silicosis, a debilitating lung disease that can be fatal. It is caused by breathing in dust particles in the mines, weakening lung capacity. Mario Gomez, one of the trapped miners, suffers from it, and his father died from it.

A VOICE FOR THE VOICELESS

It is for these reasons that the Catholic Church intervened wholeheartedly in the national debate that the accident triggered. The disaster in Atacama brought to light the unequal conditions of the Chilean people, especially in the small mining industry.

"This is a very rich region, but it has not always been rich in safety for the people," said Gaspar Quintana, the bishop of Copiapó.

The tragedy in the San José mine revealed yet again the harsh working conditions for the miners—a situation that the cleric knew well and that now brought him to level his criticism at the way production was conceived in his country.

The bishop's critical attitude was nothing new. Anytime he got the opportunity before the media or his parishioners, Bishop Quintana made the plea that labor not be just slavery by another name. "The moment has to arrive when Chile's social and political maturity permit everyone to live in a climate of dignity, and one aspect of this dignity is humane working conditions," he explained.

Bishop Quintana knew plenty about the difficulties the miners faced day in, day out. In 2001, he arrived in the area and familiarized himself with the realities in the mines—the principal economic activity of the region. His discourse focused on the intent to "humanize a life that is always on the brink of a collapse, of an accident." And this crusade gathered strength after the tragedy.

But the San José accident probably did not take the bishop completely by surprise, for he knew the conditions under which the real work of mineral extraction occurred. "What happened in the mine is a shame," he said, and now he was more adamant than ever in his criticisms of the deplorable conditions for the workers in the small and medium-size mines.

"Chile is a major mining country," he said. "Working behaviors, social behaviors, and even government and business policy have not always risen to the level of what mining means to the country. It is a job involving high risk, not always good treatment, and not even always good pay."

The bishop felt that a country about to mark its bicentennial, yet continued to allow the negligent labor practices revealed by this accident, must "learn the lessons of history." It must create an ethical conscience, "so that labor is not a case of slave owners and slaves, but a dignified profession!" he exclaimed, with the vehemence of one who has seen injustice.

His sobering words went out to a country that had been driven by good macroeconomic growth statistics but that had conveniently overlooked the cost these gains incurred in safety and quality of life for the workers.

"At times, the level of development muddles the plans," he explained. "I repeat, point-blank, that the concept of development must be driven not by the economy but rather by the treatment of the human being. We must end the pain caused by a careless, negligent, and irresponsible appetite for profit, to make more and more money while not giving even minimal safety to the workers." He said these words with a quiet fierceness.

The emergency situation in Copiapó revealed the urgency of improving conditions for the workers in the mines. For the bishop, it was also a wakeup call to the employers in the mining sector. It was now time for them, humbly and honestly, to acknowledge their errors and omissions and make things right.

"They have to pay to the degree they are responsible for all this," he said. "Not everyone is accountable to the same degree, but this has made us think about everyone. I'm not saying that in all [the businesses], but in some of them, the situation is clear. Policy regarding salaries, workplace maintenance, and safety still falls far short."

EVERYONE WANTS TO HELP

Ordinary citizens, wanting to help in the rescue effort, offered their suggestions, which often ranged from merely strange to downright harebrained.

The undersecretary of mining, Felipe Infante, between laughter and expressions of horror, revealed the contents of two letters he had received. He had to read them several times before he could believe that these were intended as rational proposals by seemingly sane people.

"One person recommended that we send a horde of rats mounted with cameras down to look for signs of life in the mine, while another man suggested that we fill the place with water so the miners would float to the surface," the undersecretary said, astounded.

But Infante got his biggest surprise—though the intentions behind it were no doubt sincere—when he received a map of the San José mine, sent from Quilpué, a little town in the Valparaíso region, by the granddaughter of the site's original owner.

"The thing is, the map went only forty meters deep, so it didn't do us any good," Infante explained. "Still, I suppose it's the thought that counts."

CHAPTER FIVE

"WE ARE OKAY IN THE REFUGE— THE 33"

August 22

Seventeen days had passed since the disaster, and it was just another day like the string of days before it: at once tranquil, expectant, and tense, since there was still no sign of life down below.

The calm atmosphere didn't lessen the underlying nervousness felt by all those living at the camp. At times they seemed to be preparing themselves emotionally for the worst. Other times, they brimmed with good humor, the most noticeable and unquenchable trait of the Chilean worker. It seemed that laughing in the midst of tragedy was a survival mechanism, and the least little thing could produce the laughter needed to boost everyone's morale.

It was right around the evening meal, which invariably began right on time—in this, the women never slacked, investing the task with their perennial care and bustling effort. People chatted in different parts of the camp, which stood a few yards from where the drilling operators, technicians, and engineers worked unstintingly, every day with greater intensity.

Among them was Nelson Flores, who just ended his shift running one of the drill rigs that were boring deep below the desert terrain of Atacama, searching for the whereabouts of the thirty-three.

Methodical as always, he stepped away from the giant rig for a moment to hoist onto a truck one of the drill sections that his team had pulled up from the depths, totally covered with dirt and mud.

Around him, his coworkers split up to clear the work area and help retrieve the rest of the drill, which had just finished another fruitless day of work.

But Nelson noticed something odd about the color of bottom of the drill section. With childlike curiosity, he went over to it and reached out his hand, curious to have a look at what, on closer inspection, proved incredible.

One part of the drill was an intense red color, which clearly didn't match the normal paint of the metal.

He stood silently, lost in thought, as if he were witnessing a revelation from on high.

It can't be, he told himself.

But this confirmed something that happened a couple of hours ago. He had felt a faint banging come through the drill bit—faint from his perspective but no doubt hard down below. Hard blows against the drill bit from some heavy tool, swung by a group of desperate workers.

It can only be them—or some of them anyway, he thought. He concluded that they must have beaten on the tip of the steel when it reached the refuge, and then managed to paint the drill bit as a way to yell up at the rig operators that they were down there, all right, and still very much alive.

Nelson Flores shuddered. He gave a loud yell to the rest of his workmates, to show them his wondrous find. They all came running, scrambling and stumbling in their haste, to gather around.

There it was, there could be no doubt: the steel drill painted by the workers they had been hunting for the past seventeen days.

The drillers grinned, they crowed, they hugged each other, they high-fived. Finally, after all that drilling, they had found the miners.

Now that they had reached their objective, the group held an emergency meeting to figure out the next steps. Time was running out, and uneasiness was running high.

Amid all the joyful reassurance and guarded happiness, one of Nelson's helpers, at some distance from the others, could not take his eyes off the red-painted drill bit that still lay on the ground, all but forgotten.

He tried to clean off the tip, to knock off a large lump of mud had come up stuck to the drill head. No one noticed him.

He stared, incredulous; his heart began beating faster than it had in the past two weeks. He looked again. Beneath the crust of mud was a clump of wet plastic bags, tied with rubber bands to the steel. His first thought was to save them as souvenirs, but then he did a double take, for bound up in the clump was a wad of paper.

Carefully, with the delicacy of an archeologist uncovering an ancient manuscript, he opened the plastic bags and undid the rubber bands, his heart thumping wildly. With that mixture of fear and uneasy hope that little children feel at Christmas, he worried loose the plastic bags, taking his time, and then just stood there gawking. He was looking at a man's handwriting . . . yes, writing.

"There's a message here!" he said aloud, though in all the hubbub around him, no one heard.

It was a crumpled piece of paper—an ordinary page torn from one of those gridded notebooks that schoolchildren use for their mathematics classes—written on in red ink. The shock froze him. He could not believe the words that appeared before his eyes.

"*We are okay in the refuge—the 33,*" read the wadded page, as wrinkled and dirty as any other of the millions that got tossed in trash bins every day—but this one bore the most momentous, most awaited, most dreamed-of message since August 5. The sentence was short, yet it hung there in his consciousness endlessly.

"They're alive!" he gasped. *"Alive!"*

He thundered the words, and the wind helped out, carrying the astounding news to the small group of workers who were still coming up with a rescue plan. Everything froze. Even the dust seemed to stop moving in the air, as if every particle were stopping to bear witness, and nature itself were echoing the human sense of wonder and relief. The drill operators lost their composure, their elation turning them into children, and they cried with an abandon that men seldom permit themselves.

The man who made the discovery looked in all directions, thunderstruck, unwilling to accept the miracle. It was true that the thirty-three were alive. "Thank you, God," were the only words going around in his head. Without another thought, he ran down the hill to the family members, to give them the joyful news. But one of his mates stopped him. He reminded him that they could do nothing without the consent of the authorities.

"But we *can't wait* till the president arrives!" he replied, stammering every syllable. "We have to tell the families—they're living with this anguish, and their husbands, brothers, and sons are alive!"

Even thought the operators had been ordered to give every bit of information to the government first and only then to the families, this overjoyed man, recognizing his right and, indeed, his duty, broke every protocol and pelted down the hill as fast as his legs could carry him, down to Camp Hope.

There was confusion among the families. The news was too powerful, too potent, and no one was prepared to hear it. The press tried to get information. The spokesman for the ministry of mining shouted from the roof of a pickup truck to the journalists, demanding that they say nothing.

"We can't let the news out so irresponsibly," he said. "We have to confirm with the authorities and wait till the president arrives."

But the genie was out of the bottle. The reporters, perhaps with more guts than sense, ran with the bombshell: the wondrous news that the miners of Atacama were alive in the refuge of the San José mine flew around the world, and the technical details be damned.

The crumpled piece of paper stated clearly that the thirty-three trapped men were alive. Minister of Mining Laurence Golborne, with his ever-present red jacket, smudged and dusty now from the days at the site, and his lock of hair that kept falling down to his eyes, grabbed his cell phone with trembling hands and called the president. With a steady voice, nonetheless full with emotion, he recounted the marvelous news.

Golborne knew that, starting today, the story changed. He waited for the president's arrival before making it official. It was necessary, he explained, because the news was too important and the state had to take responsibility before the families, Chilean society, and the international community, which had been closely following the events in the north of the country.

Out there in the desert, emotions exploded into euphoria. The news traveled like a lit fuse. It seemed too wonderful to believe: *We are okay in the refuge—the 33.* The families were happy, thrilled. They cried and jumped up and down like children. Now they knew that their faith and their prayers to God and the Virgin had worked. The ferocious hugs continued, tears flowed at every encounter, and euphoria filled the national and world press, so that any human commentary was only so much white noise. Tears dampened faces and clothing and streaked makeup; excited hugs moved in waves, as if wanting to bring everyone together. Everyone at Camp Hope—the wives, children, rescue operators, politicians from the administrations, representatives from the mine, cooks, clergy, vendors, and journalists—all of a sudden their differences were gone. They clung to each other with a force they had never known before. Some fell on the ground, as if seeking the embrace of the earth itself, as a repository for this surfeit of joy that could not possibly be contained.

Teapots boiled dry and biscuits burned while everyone ran about, unable to sit still and wait for more news. "They're alive!" shouted the wives. Cell phones rang on all sides. Everyone was calling their families, so excited their shaking fingers had trouble pressing the numbers.

Hours Later

The president arrived, and the press waited eagerly on an improvised platform near the little shrines the families had made for their men. The reporters practically climbed over each other to get the best vantage point. Camera crews and microphone booms rushed together, bulbs flashing to register the historic moment. Golborne tried to compose himself—a virtual impossibility in the moment.

The press calmed down a notch as President Sebastián Piñera held up the note from the bottom of the mine in his hands. Brimming with happiness in his starched white shirt, he announced to the world that the thirty-three miners were alive.

"This came today from the bowels of the mountain, from the very depths of this mine," said the jubilant leader, filled with a euphoria never before seen in his long public life. "It is the message from our miners, telling us they are alive, united, and yearning to see the light of the sun and embrace their families. I want to say that all Chile is weeping with emotion. But even more, I want to thank the miners for holding out two and a half weeks, alone, and I want to thank the families, who have never lost hope, and to thank this entire rescue team who were so unstinting in their efforts. The news fills us with happiness, with strength. I feel prouder than ever to be Chilean and to be the president of Chile. *¡Viva Chile, mierda!*" he concluded in obvious ecstasy.

For Piñera, these heartfelt words were the perfect beginning for September's Homeland Month. Better yet, this was also the nation's bicentennial.

"But now we must continue our efforts," the president added. "We have to line the hole, we have to provide them with food, lighting, communications, but the most valuable thing is already here: moral support."

The information spread instantly around the world. In every part of the country, the people came out to celebrate, most of all in

Copiapó, the city where most of the miners lived and that had been in mourning since the collapse.

The black flags that hung from every house to convey the inhabitants' grief were fast disappearing, for they now had no reason to be there.

Joy lit the face of every Copiapino and of all Chile.

In Santiago, meanwhile, auto horns blared just as they do when the Chilean national soccer team scores a victory. People gathered in the vicinity of the landmark Plaza Italia to celebrate.

The information pulled at the heartstrings, not only of the miners' families and the rescuers, but of all Chileans who had been worried sick over the fate of their countrymen. The Church did its part with religious services outside the mine.

Hours later, the minister of health, Jaime Mañalich, arrived at the site. He knew that now, after the discovery of the famous piece of paper, his role was even more important, since he had to keep the miners alive and in good medical condition, whatever it took.

"According to what we know, no one has any serious injuries," Mañalich announced.

He also knew that starting now, the families were an even greater concern because now they were agitated and expected a quick rescue. He acted immediately, explaining that it was necessary to calm the anxieties of those aboveground as well as those below.

"It is urgent that we determine the psychological condition the miners are in, and learn the physical consequences of spending so much time without getting normal nutrition. They need to understand that it will be many weeks yet before they see the light of day. We need to do a diagnosis, explain to them the situation. There are psychologists on site. We need to take the lead and support and prepare them for what lies ahead, which is no small thing. We need to give them a realistic assessment of the order and timing—and the uncertainties—of the rescue task still to come." Mañalich reassured the families that if the miners weren't physically hurt and if the rescue operation could deliver food to them, the men could hold out for weeks or months if they had to, until the rescue could be carried out safely.

The press and the families asked repeatedly what the next step was. Mañalich—still in his red government jacket rather than his white medical smock—replied that they needed to send down water quickly. Then they needed to work up a diagnosis of each man's current state of health.

The rescue crews returned to work. Now that they knew the location of the miners, they needed to figure out how to deliver food and supplies to them. The medical team especially needed to get them hydrating water and salts, to be given during the first six hours once they could make contact via the borehole.

Mañalich, ever the strategist, explained things in terms of three concentric decision rings. The first ring of decisions involved attending to the miners and the rescuers. The second, wider circle involved the family members in the camp, who would surely be growing in numbers now, and the medical team needed to take care that no disease or infection appeared. The third circle extended outward to the hospital at Copiapó, where intensive care units were readied, a blood bank was equipped, and surgeons were alerted and placed on call. And eventually, if necessary, they had the support of Asociación Chilena de Seguridad.

The government now projected the rescue for sometime before Christmas. And the press, the rescue workers, and the families knew that this episode was unprecedented in Chile's history: In the long history of mining accidents, there had never been a rescue of so many miners, trapped for so long, so deep beneath the surface.

Now that the medical experts on the scene knew that the miners were alive, they had to understand what their bodies had been exposed to in the past seventeen days, and what they would be contending with in the weeks to come. Almost a mile underground, the mine was a humid and dank place. It was over 90 degrees there morning and day. The precipitation created a breeding ground for fungi, bacteria, and plenty of other pathogens: The doctors braced themselves for respiratory ailments and worse.

While the euphoria grew in Camp Hope—which now more than ever deserved its name—the specialists continued their work.

They began to lower the *palomas* (plastic capsules) down to the bottom of the mine once again, this time loaded with medications, serums, and tetanus vaccines. Mañalich optimistically asserted that with a little luck, they could bring out all thirty-three alive and with no collateral illness.

Hours Later

There was more than the single wadded piece of paper in the clump of plastic bags attached to the drill head. The message that brought such joy to the Chileans hadn't arrived alone. On the head of the drill rig that had bored down to 2,257 feet was also a letter from the miner Mario Gómez to his family.

President Piñera was beside himself as he delivered to the press the message of life from the thirty-three. Glowing with emotion, he read Gómez's letter:

Dear Lila, I am fine, thanks to God. I hope to be out of here soon, patience and faith. . . . God is great. We are going to get out with the help of my God. . . . Even if we have to wait months for the chance to talk, I want to talk with Alonso [son-in-law]. This company has to modernize. I want to say to everyone that I am fine and that I am sure that we will get out alive. Okay, Lila, I'll see you soon. We will stay in touch. The first borehole has broken through; now we have communication. A little water trickles down one wall here. . . . We're in the refuge. . . . May God guide you, and greetings to my family, I love you. Mario Gómez.

His wife, Lilian Ramírez, swelled with happiness. Her husband was highly experienced as a miner but was also adept at dealing with the elements. Not only had he worked in this mine, but he often camped outside, with nothing but cardboard to cover him. So, the young woman stammered, he was strong and was not about to let his companions down.

Lilian was visibly exalted receiving the note from her husband. The other wives, too, were walking on air. Mario's words were not just for Lila. There could be no doubt that they were also for every one of the family members listening so attentively to Mario's words in the voice of the president. Happiness spread throughout the camp, and the peace of mind at knowing that the miners were all right spread across the world.

After this letter, the mood among the family members changed. Now they knew that their loved ones were alive, but they didn't know for certain what shape they were in, let alone how long they had to wait to see and embrace them once again. Faith prevailed, but traces of doubt remained.

THE DEFINITIVE RESCUE PLAN

The day after the discovery, the minister of mining, Laurence Golborne, awoke feeling pleased, satisfied with the results of the drilling, and he suspected that the coming days could be even more favorable. Golborne had the absolute conviction that proof of the miner's lives signaled a turning point that boded well for the future.

With his brow furrowed against the northern Chilean wind, the minister said that the miners had been tremendously adept and smart. Now for the most important part: the race against time to bring the thirty-three to the surface alive. Golborne met with the experts. He received a report on the progress being made, the technologies being used, and the companies providing services. The reports came in around the clock and were always sent on to President Piñera.

However, amid all the happiness, Golborne kept his head down. He was a realist. Beyond staying in contact with the miners and beginning with the definitive rescue plan, he kept one eye on the long haul, which, according to those in the know, could take anywhere from three to four months. Now patience would have to show its colors.

The rescue news did not go well with the families. They wanted their men on the surface *now*. There was anxiety in the air. The projected three to four months struck them as too long, and they did not want to keep waiting for their loved ones. Nevertheless, Golborne assured everyone that the rescuers would pursue the fastest—but also the safest—option.

MARÍA, THE MAYORESS

"We will pray with you!"

The cry resounded over the boundless desert waste and made its way forever into the hearts of the women. Living each day with fear and fragile hope, they knew precious little of what was happening to their husbands, sons, brothers, lovers, and friends, who had managed to send to the surface, from 2,257 feet underground, the most important message of their lives.

A shriek came from the mouth of María Segovia, the sister of Darío Segovia.

María, a seller of meat pies at a public market, left everything in the far northern city of Antofagasta and came to Copiapó, not just to get a better idea of what was going on at the site, but also to see that everyone fulfilled their responsibilities in the rescue effort. Already her strong personality, faith, and sheer obstinacy had turned her into the leader and spokesperson for all the women in the camp.

She lightened their anxiety and dark moods, calmed their anger, empowered them, interceded with the authorities in charge of the rescue, and organized Camp Hope, which had experienced a growth spurt in recent days.

Because of this and much more, María was called *la alcaldesa*— "the mayoress."

The media sought her out.

She never stopped, even while keeping her deepest fears to herself, so as not to bring down the others.

She raised her voice for the families and pitched a tent at the entrance to the camp. That tent became her home, with the Chilean tricolor always flapping in the wind.

María's new home was cozy and hospitable, prepared for conversations with family members and reporters: She had hot water, bread, and tea—and the inevitable maté gourd, which she shared at times with Laurence Golborne himself.

The drinking of maté is a singular custom. Maté has its roots in southern Brazil, Uruguay, Argentina, and southern Chile, and partaking means much more than merely drinking an infusion of a slightly bitter, energizing herb.

The tradition consists of sharing the same *bombilla*, or drinking straw, among the entire group as an undeniable symbol of communion, loyalty, commitment, and complete confidence in the other—a ritual far from the prejudices and separation that typically pervade contemporary life.

And so it was this night in Camp Hope. The uncountable stars of the firmament witnessed the circle that formed around the little stove for the maté ceremony—a ceremony that brought together families, rescue workers, reporters, authorities, and even the merely curious who came to be part of one of the most publicized disasters in human history.

"What do they do here, those miners?" It was a variation on the theme of the folk group Illapu, sung time and time again around the Segovia family campfire. They occupied the first big tent at the entrance to the camp, and they were the loudest group in the camp.

"I came here the first day, when there was no one, and sat down on a rock to await the news," recalled María Segovia.

Like her, there were many others who had not budged from Camp Hope.

"Now it's my home," she said proudly.

CHAPTER SIX

THREE PLANS

HOW DID THE MEN SURVIVE?

Even though the miners vowed not to talk about their ordeal in the mine before they were discovered alive, some details have come to light about how they survived those first dark days, and how the group reacted when the first drill broke through. The shift foreman, fifty-four-year-old Luis Urzúa, said in an exclusive interview for this book that the drill was received by the men like a God sent down from heaven above, after seventeen long, desperate, agonizing days of uncertainty. That drill represented hope, optimism, and faith.

They soon set to work affixing messages scrawled on pieces of paper to the drill. One said, "Send me some food." Another read, "I'm hungry." "There were lots of scraps of paper," Urzúa recalled, "but God wanted only the ones that had to reach the surface to get there." And there was the most important message of all, that confirmed that all thirty-three men were alive: All thirty-three are ok in the shelter.

Urzúa remembered that for the first few days following the collapse, they had "very little food," and by the time they were found to be alive seventeen days later, they were only eating once every forty-eight hours, to try and always keep something in reserve for later. Their diet consisted of a teaspoon of tuna fish, a quarter of a canned peach, and a sip of milk that eventually spoiled. The men were so self-disciplined, they managed to survive on a supply of emergency rations that was meant to last for only two days. They even had some left over when they were found. They had managed to survive on a diet of only 150 calories per day on average.

The potable water ran out very quickly. Still acting as the shift leader, Urzúa had the men work on digging wells, to try and find more water. This one action is widely credited as being the key to the men's survival for those first seventeen days. After a well had been dug, someone was assigned to drink from it, to find out if it was safe. They were monitored for any adverse physical reactions. "Some of the men started to feel really sick, they had terrible stomach cramps, but there was nothing we could do, they just had to endure it," Urzúa remembered. But this method allowed them to discover which of the wells was potable.

The men spent most of their time in the small emergency shelter space, about 550 square feet, because they knew that was where rescue teams would look for them. There was no light in the shelter, so the men used the lamps on their helmets, very sparingly. They had access to machines equipped with lights used in mining operations, but their engines would have polluted the air supply so they couldn't use them. From the reinforced shelter, they could still reach other access tunnels and ramps farther out in the mine. But they weren't stable; more rocks had fallen after the initial collapse, so the men usually didn't wander too far. They did use a separate chamber away from their shelter as a bathroom.

It was so hot, the men were half-naked all the time. "Imagine what it's like to have to live in 40-degree Celsius [104 degrees Fahrenheit] heat. Imagine living in almost 100 percent humidity," Urzúa said. It would have been enough to kill almost anyone.

THE PLAN TO DRILL THROUGH ROCK

"Ladies and gentlemen, thank you for coming," Laurence Golborne said with his usual impeccable manners, greeting the dozens of reporters before him.

It was a cold Sunday at the mine, well after eight o'clock at night and quite dark.

It was a special occasion. For the first time since the collapse, the reporters could cross the barrier that separated them from the camp. Now they squeezed in however they could under the big white tent where the authorities met daily with the families. The place was filled with more cameras and microphones than air to breathe.

The journalists waited impatiently for the minister's words. Golborne had just gotten off the airplane that brought him from Santiago and made his way straight to the mine. He looked as though he had something important to say.

"We are going to present the three different machines that will do the drilling," he explained. "We view it as a sort of healthy competition between them to get to the miners as fast as possible."

With equal parts nervousness and enthusiasm, he gestured toward the PowerPoint image that the families had been looking at each day, figures and drawings informing them of the status of the operation.

Golborne and engineer André Sougarret went over the three proposals they had received. The moment the presentation began, the room fell silent as everyone gave them their full attention, not wanting to miss a single detail.

PLAN A

The minister explained that the Raise Borer Strata 950 was a machine used by Codelco in its Andean division to bore mine shafts and ventilation ducts up to twenty-six feet in diameter.

It was tasked with drilling the rescue hole using a mechanism called a "trepan," which could reach a depth of 2,624 feet—the distance needed. After drilling, a second pass, with reaming equipment, would widen the access hole from almost thirteen feet to almost twenty-six feet diameter—wide enough to accommodate the rescue cage.

Installed in the spot nearest the mine entrance, this machine would be the first of the three to start work and would average forty-nine feet a day.

With all the inevitable adjustments, the engineer's calculations had them reaching the goal in mid-October—a good possibility.

PLAN B

Now it was time to see what the Schramm T–130, the machine that had an eighty-five-foot head start, could do.

Its tower was visible from anywhere in camp, and the noise of its motor was the background music for the daily routines of the families during their ongoing vigil.

The gigantic mass of metal, owned by U.S. firm Geotec Boyles Brothers, came from the Collahuasi mining company, located in Antofagasta. It was operated by two North American experts who had been working with a similar rig, drilling for water in Afghanistan, when the tragedy happened at the San José mine. All present nodded their agreement that this plan, too, sounded good, and immediately, Jeff Hart and Matt Staffel were brought to Atacama.

The T–130, mounted on a heavy-duty truck, was commonly used to drill water wells and was able to reach a depth of 3,280 feet. The immense apparatus widened the space from over eleven feet to twenty-seven and a half feet in diameter using the "Down the Hole," a type of drilling hammer that combined five heads in one. The T–130, from Canada, had not yet been used in Chile, so no one knew how it would perform in the extremely hard rock of the San José site.

PLAN C

The RIG 422, as it was known by the experts, came from Iquique, in the far north of Chile. Forty-two trucks brought the steel sections to build the tall tower and install the heaviest weight that the San José terrain had ever borne.

It was a slender construction 147 feet high, used in petroleum and gas exploration. With its technique using a tricone drilling hammer, it was the fastest of the three drill rigs. The RIG 422 would drill thirty-five inches in diameter for the first 164 feet. From that point, the shaft would narrow to twenty-seven inches. Its projected time to reach the miners was also a little less than two months.

Much of the RIG 422's efficiency depended on geological conditions, on the time it took to assemble and install, on the large work area it required, and on the uncertainty concerning the type of rock. This machine was very fast in soft terrain, such as the Amazon basin, where it was usually used to drill for oil, but no one knew whether it would work as well in a place this dry.

The audience was hesitant. Some leaned toward one of the three; others were skeptical about the whole process. Everyone quieted down when they heard Golborne's quip as he finished his presentation:

"Place your bets, gentlemen."

Sougarret laughed and said that he already had. (Privately, he said his chips were on plan B: the Schramm T–130 from Collahuasi.) The minister, obliged to reveal his preference, went for option C.

"I don't care if I win the bet," he added. "What matters is that one of the machines completes its mission in the least time possible."

A NEW TOY

The work of getting help to the miners proceeded at full throttle. Officials lowered a video camera along with the medical supplies and drinks. Another arduous task. Cables, electronic components, and other small gear began going down, one capsule at a time, to reach the confined workers.

After an operation involving much delicate work, the miners finally had a tiny lens installed at the bottom of the mine and connected to the surface—a link that could now transmit live what was happening inside the shaft.

Everyone wanted to see the faces of the cloistered workers, to see how they were and how the place looked, to know more details about what life was like down in the dark and the damp that had been their constant companions for more than two weeks.

Although the transmission was instantaneous, the images first had to be checked out by the government authorities before being broadcast, to make sure nothing would portray the miners in an undignified or embarrassing light.

Meanwhile, Golborne, Piñera, and the rescuers waited eagerly before the laptop computer that would show them the faces of the buried men. Beside them sat the miners' next of kin—privileged spectators in a place restricted to them alone.

Deep in the earth, a group of miners, with their lamps, drew near the video camera like children seeing a new toy. Above-ground, the group saw little lights approach on the screen. It really was the miners—the first of the images from 2,296 feet underground.

One of the miners, Florencio Antonio Ávalos Silva, approached the screen. His family recognized him at once. "It's Florencio . . . it's him!" they shrieked on seeing his face. The thirty-one-year-old worker, married with two children, was in the refuge along with his brother, Renán Anselmo.

In the little room of spectators on the surface, this image generated not only elation but also plenty of confusion as everyone tried to figure out who the miner was. With each face that appeared, a dozen people were sure they were seeing their long-lost loved one.

Early on, the parents of Jimmy Sánchez, who was nineteen years old and the youngest in the group, were sure they saw their son there on the screen, but moments later, other families were sure it was *their* son, husband, or father, so confusion reigned much of the time. The miners, meanwhile, unaware that everyone up above was trying to recognize them, waved their arms and gave signs that they were alive and in good spirits.

After about five minutes, President Piñera went out to find the press who were posted at the camp, to tell them about what the families were seeing. He assured the reporters that he had seen the workers with his own eyes through the video cameras.

"I saw eight or nine of them waving their arms," he said. "They had their lamps on and were apparently in good physical condition. They recognized the camera and signaled that they were happy, but we couldn't make auditory contact with them."

And indeed they couldn't, for it sounded as if a nearby waterfall was drowning out their voices. Quickly the authorities went to work, trying to get more than just visual communication. They needed to hear from the miners' own lips that they were well.

The miners appeared in relatively good heath, but the medical team was able to get a more detailed picture from looking at some of the video images taken. They could see that some of the miners were suffering physically.

The medical team was very concerned about the emaciated appearance of some of the men. They were also worried about the men's environment, since they were basically living in a hole in the ground, like burrowing animals.

The possible long-term effects on the miners' health from the dangerously high levels of humidity in the air was cause for concern. The medical team was also focused on the lack of natural light and well-oxygenated air.

Health Minister Mañalich explained that they were concerned about the possibility of a serious epidemic breaking out among the men. They had to be very careful not to introduce any disturbances in the environment through the *palomas,* the small torpedo-like canisters that delivered the supplies.

As part of their strategy of preventing infection, the medical team sent down thermal clothing and anti-fungal socks made from a special material with copper threads, and a cleanser and ointment for the men's irritated skin. In such a high-humidity environment, conditions that are usually just minor annoyances, like foot fungus, can turn into very serious infections, since there is no way for them to dry out. Even small cuts and abrasions could develop into deadly infections. Unable to take showers, the men were also vulnerable to an infection and inflammation of the hair follicles, causing constant itchiness.

The supplies were delivered through their small little tube, stretching 2,300 feet below the ground. At first it took quite a while for the canisters to make the journey, so supplies were slow in coming.

The most important thing at first was to get drinking water to the men. Then, they had to be fed. But people who have been starving must begin eating again in a very careful way. If not, they could suffer from refeeding syndrome, resulting in dangerous drops in the levels of phosphate and potassium in the body. In a starvation state, the body does not produce insulin at the same rate it normally would, and a sudden intake of carbohydrates can cause a radical shift in insulin production, with potentially life-threatening consequences.

The miner's caloric intake was increased gradually. At first they received 1,200 calories a day, and several days later that was increased to 2,000 calories a day, which would allow them to maintain their health. They also had to be mindful of not gaining too much weight, since they would have to fit into the very narrow Phoenix capsule to be rescued.

The men were also immunized against tetanus, diphtheria pneumococcus, and influenza.

The rescue team was able to pump in compressed air through an air hose snaked down one of the narrow boreholes. This action was extremely important, since the men had been breathing with increasing difficulty. Also, it would hopefully help to counteract the extreme temperature. The deeper a mine is, the hotter the atmosphere, and the mines in northern Chile are especially deep.

The prolonged absence of natural light affects sleep cycles and can also impact mood and cause a deficiency in vitamin D. This would have an effect on the miners' emotional health.

Absence of light can lead to conditions like seasonal affective disorder, which is a type of depression. It also interrupts a person's natural circadian rhythm, which tells the body when it is time to wake up, and when it is time to go to sleep. The medical team monitoring the miners was concerned that the lack of light would lead to feelings of depression and cause fatigue.

Establishing good hygiene was also a concern. The doctors advised the miners on how to keep their small living space in the shelter clean. One of the first orders given to the men was to dig a latrine in the shelter, partly to give them something to do and keep them busy. The accumulated waste would produce an odor that over time the men would get used to and tolerate just fine, said Dr. Jorge Diaz, one of the medical team monitoring the men. "If we went down there from the surface, the odor would be overwhelming, but the body can adjust itself to this kind of sensory input."

CHAPTER SEVEN

"THE MINE HERE"

N ow appeared the figure of Pedro Gallo, inventor of the Gallófono—the telephone that would allow the rescuers to communicate with the trapped miners. Amid the happiness and celebrations in the camp, Gallo prepared for a complicated initial attempt. He had already been ignored earlier in the rescue by the engineers from Codelco, who couldn't see the usefulness of his invention, and he wanted no surprises.

The moment had arrived when he must show everyone that all his work was worth the effort. And the device needed to function perfectly when those in charge of the rescue used it to make the first telephone contact with the miners.

The device was lowered inside a capsule almost twenty feet long. "It fit in perfectly, with a millimeter to spare on either side," said Gallo.

They sent it down the duct—the life-giving umbilical channel that now united the workers with the outside world. And an hour and forty-five minutes later, the men 2,296 feet below the surface connected the cable to the Gallófono just as if it were a conventional telephone.

Up above, Gallo connected "2" on the telephone switchboard—from this point on, it would be the number identifying the team underground.

Minister Golborne picked up the handset.

"Hello? Can you hear me?"

The answer took a second to arrive, accompanied by a faint metallic buzz.

"The mine here."

The voice was that of the shift foreman, Luis Urzúa, and the buzz that Gallo heard was the sound of the thirty-three discussing who should make their first spoken contact with the outside world.

"We're fine here," Urzúa reassured the minister with his first words, "just waiting for them to rescue us."

Golborne could scarcely believe his ears as the voice underground went on with its story.

"Well, we've been drinking some water. But we haven't eaten much lately—all that we had in the refuge."

The minister of mining, a little calmer now but deeply moved, told Urzúa that all Chile had been following the drilling.

"This has been an effort by the whole country, which has been watching the search-and-rescue process since it began," he said. "Be assured that you are not alone. Yesterday, all of Chile celebrated in the streets throughout the country—overjoyed that we had made contact with you. Today they are going to be even happier to know that we have spoken with you."

But Urzúa was less interested in the excitement that all Chile was feeling than in the fate of a coworker who had been leaving the mine on the day of the collapse.

"We don't know if he got out or not," he told the minister.

"Everyone got out unharmed. There is not a single death to mourn," Golborne said. This prompted loud applause and cheers from the trapped workers, who had been more worried about their comrade's fate than about their own misfortunes.

"I want you to know that here outside the mine a camp has been set up, where all your families are staying," the minister of mining continued, inspiring still more rejoicing among the workers. "You are not alone. Your wives, children, and families have been here with you since the first day. You can be utterly confident that we are doing everything possible to bring you up here as soon as we possibly can. A great hug from all of Chile, because all of Chile is with you."

Not yet ready to change the topic from the fateful day when they were buried, Urzúa told the minister that they looked for an escape route through the ducts, but since they didn't have ladders, they couldn't search high up. Hearing this, Golborne told them at once, "Those shafts are blocked . . . it is most important that you stay away from the rock in that area—it is extremely unstable."

This conversation between these two men, between the surface and the bottom of the shaft, set off a little festival on the surface. Everyone yelled and hugged.

Pedro Gallo was the only one who did not join the group. After hearing the conversation, he heaved a deep sigh of relief, walked a little way off, and lit a cigarette. His eyes filled with tears as he looked on, moved by the scene of jubilation.

BLESSED TELEPHONE

From that day on, Pedro Gallo conversed daily with the miners to discuss their needs.

"I talk with them every day, from morning till night," Gallo recounted. "And I stay up all night with them. I've become a psychologist, counselor, friend, courier, and love messenger for the miners."

Gallo's unimpressive-looking little device was becoming the bridge that would keep up the hope of the anxious miners and their families. And indeed, the Gallófono would continue to be used even at those times when the video camera wasn't working.

It was through the telephone that the outside world got its first impression of the state of the thirty-three. And it allowed the doctors on the surface to guide Yonni Barrios in diagnosing the various complaints of his comrades.

The first major complication was the supposed appendicitis of one of the workers, which turned out to be a false alarm—corroborated thanks to the veritable medical encyclopedia of doctors on the telephone line.

On another occasion, a dentist got on the phone to give Yonni instructions for an emergency tooth extraction with no anesthesia.

"Do you have any experience pulling teeth, Yonni?" he asked.

"One time I had a molar that was hurting, and I pulled it out with a pair of pliers," the miner replied naively.

The operation stopped at that point. Faced with Yonni's apparent lack of necessary sensitivity, the medical team decided to treat the tooth with antibiotics.

The Gallófono also served as a guide for technical conversations instructing the miners in tasks key to the rescue operation. All the conversations were carefully managed by the rescuers, always with a view to making a difficult situation more bearable.

Down in the refuge, from the very outset, Pedro Cortez and Ariel Ticona took over the technical aspects of the communications. They took turns caring for the equipment and keeping it functioning.

"They have given themselves the title of manager and assistant manager of communications," said Gallo.

From the surface, they got instructions for assembling the systems that were lowered in the capsules.

Luis Felipe Mujica, another of the rescue coordinators, recalled that when they installed the conduit through which the fiber optics and the telephone cable were lowered, the miners had to assemble the part of the videoconference system to be installed inside the mine.

Down in the capsules went the small projectors, cameras, microphones, and other gear, even including a blackout curtain. But before the engineers up above could tell Cortez and Ticoma how to assemble the system, they had part of it already set up purely by instinct.

NIX ON THE DIRECT RESCUE

As the initial exhilaration began to dissipate, the camp grew calm while the authorities concentrated harder than ever on the rescue.

The next steps were clear and concrete. The fragility of the mine prevented a direct exit; therefore, according to the experts, the boreholes to free the miners would require two to three months to complete. Meanwhile, the men must survive as they had until now, in the refuge located in the very bottom of the mine.

It was a place totally lacking in comforts, without even the most basic conditions for a minimal existence. The darkness and dampness, the confining closeness, and the lack of food made it a hellish place to live.

The refuge in the San José mine was a sort of giant crack located in one of the almost five miles of passage that curved in a spiral down to the bottom of the mine, nearly 2,296 feet below sea level.

The space, almost 165 square feet, in which fifty people could fit, was supplied with small benches and, in theory, enough oxygen tanks, food, and bottles of water to withstand a confinement—in theory because, according to the first accounts of the miners themselves, at the time they were trapped there was little food.

Including the accessible mine tunnels, they had a space of a little over a square mile to move about in, and the refuge was connected to a gallery of 656 square feet. The cracks in the mine, an old excavation from the late nineteenth century, allowed enough air in for them to breathe. There were batteries in the refuge, and they took turns using them to optimize the available time for artificial light. For illumination, they also used the trucks and pickups that remained in the corridor.

A DAY IN THE BOTTOM OF THE MINE

Normality—that was what the team of physicians and psychologists working full-time with the miners looked for. To avoid unnecessary anxiety and potential psychological relapses, they did not want the thirty-three to dedicate their every waking hour to thinking about the rescue. They wanted the miners to have as normal a life as was possible nearly half a mile below the feet of other mortals.

The workers' daily routine began at 7:30 A.M. with breakfast: a bottle of protein concentrate. Then came personal activities such as writing letters or taking care of some unfinished task, making beds, and so on. At midday they had lunch, then prayed and reflected as a group. This was not an activity imposed by the doctors and therapists; the miners chose to do it on their own—perhaps because they needed it.

Now there was something else to celebrate: they had hot water for yerba maté or tea at five in the afternoon, the hour of the unbreakable Chilean habit of taking *once,* in the style of English afternoon tea.

Folk history has it that the word *"once"* ("eleven") comes from the custom of the saltpeter mine workers at the end of the nineteenth century, who accompanied their afternoon snack with a slug of *aguardiente* liquor. But because of restrictions on the consumption of alcohol, they called the snack *"once,"* for the number of letters in the word "aguardiente."

Despite all the efforts from outside, some of the miners, being human, complained about the small portions of food they had been given. Their diet varied between 2,000 and 2,500 calories a day of hot meals, with scheduled times for breakfast, lunch, *once,* and dinner, and aside from that, a light meal for the night shift.

"They want to starve us to death," Darío Segovia wrote in one of his letters. In a way, his stomach was correct—the portions were small compared to normal. But according to Dr. Díaz, it was the right amount for how they were living.

Every day, a pickup truck arrived at the site, driven by Nelly Galeb, a caterer from Copiapó, who prepared the rations and brought them, hot, to the mine.

"Today we're having rice with meat," said Nelly, who was always in a rush to make sure the food got eaten as soon as possible. "And for dessert, quince jelly."

It seemed like just an ordinary social event, because the silverware, napkins, salt, and even the paper plates made the trip down in the capsules that carried the menu of the day.

"Tomorrow, for the first time, we're going to send them a plate with beans," the caterer said happily. Beans with pumpkin and spaghetti were a traditional Chilean food that the hungry guests below had been asking for, and Nelly, diligent as always, had gotten medical permission to grant their wish.

The psychologist from the Asociación Chilena de Seguridad and the coordinator of the equally well-known Operación San Lorenzo (Operation St. Lawrence), Alberto Zamora, said that every day the miners had a work routine and got as much of a physical workout as their food and energy permitted. They received the capsules and provisioned themselves; some did nursing tasks; others improved their living quarters. Everyone had something to do; everyone had an important task so that the group could continue bettering its condition.

But it was also true that the spirits of the thirty-three were flagging, and they expressed this in their letters to their families. That was the case with Jimmy Sánchez, just nineteen years old, the baby of the group.

"Since we found out they're alive, it's been a tremendous joy for everyone," said Luis Ávalos, one of Jimmy's relatives. "But from there on, they haven't been at peace, because where they are is no way for a human being to live. In his letters, he has expressed that he has to take heart, and we have told him to have faith in God because there is only a little way to go. But he's bored—as you can imagine, because he's young."

The job assignments and work shifts improved the mood within the group of miners. Antenor, the father of Carlos Barrios, thought so.

"Carlos was sleeping very little," he said, "only three or four hours a day. Now, though, with the work, he's sleeping a bit more.

According to Antenor, confinement in the mine also gave rise to a certain amount of friction and boredom from inactivity, but with the chores and assigned exercises, the situation was improving.

UNDERGROUND, A VIDEO DIRECTOR IS BORN

Officials sent a video camera down through the borehole used for lowering food, so the miners could show the outside world the place where they were sheltered and what their life underground was like.

The job of designated video reporter fell to Darío Segovia. He was in charge of interviewing the miners and also of motivating them to say something to their families. Instinctively, he knew that he and his comrades would need to make just as much of a concerted effort as everyone aboveground to get them all through to the final day of the rescue.

The world outside anxiously awaited the images. The result of Darío Segovia's work was a forty-five-minute video of history in the making that immediately began to make its way around the world.

Officials had set up a gigantic screen on what, before the collapse, had been the entrance to the mine. And there they played the video to the families, the press, and the entire world.

Spellbound, the families of the workers watched the images. They were captivated by the appearance of the faces and the bodies, and disturbed by the darkness that blanketed everything. Tears and emotion overcame those watching. Even the press were moved, and Minister Golborne's face was perhaps the most eloquent expression of the feelings that overwhelmed the watchers there at Camp Hope.

The images revealed a dark, damp, utterly inhospitable place. And yet, the space was shipshape, tidy, everything in its place.

"Here we have everything well organized," said the narrator. "Over here we have dominoes, which we made with our own hands, using material that we found here underground. This is the place where we entertain ourselves." The camera advanced and panned around. "Here we hold a meeting every day, plan our work, and divide up the jobs." The camera continued advancing slowly. "And over here, we pray."

News anchor/narrator Segovia continued his video documentary: "Here"—he held up a bottle of water—"we have a place for washing up and brushing our teeth."

The thirty-three did not look their best. They were beginning to show just how much their confinement was wearing on them. They were bearded and gaunt, with bags under their eyes after so many days without eating, and yet, despite everything, they appeared to be upbeat.

With the video camera in hand, they greeted their families, excited that they could now communicate. They showed the world what this place that had become their home was like. With pride, they showed off the improvements they had made for their own survival and how they whiled away the hours until they could at last be rescued.

At the end of the recording, the thirty-three workers sang a stirring chorus of the national anthem and then shouted, *"¡Viva Chile! ¡Vivan los mineros!"*—"Long live Chile! Long live the miners!" After that finale, there were few dry eyes in the audience.

PUZZLES AND CONDORITO

"Oratory? What on earth for?"

This was the foreign reporter's surprised response during an interview with Alejandro Pino, the regional director of the Asociación Chilena de Seguridad.

"Yes, my friend, that's precisely what I said: we sent thirty-three books of oratory to the miners," Pino replied to the skeptical reporter who was covering the news of the collapse. The man could not understand the importance of sending a bunch of trapped miners books on how to speak with poise and eloquence.

It was a sunny morning at the camp. The book had been written by Pino himself, a former radio newsman who knew a thing or two about disasters.

"I was assigned to the rescue of the other miners in Illapel [northern Chile], in the 1970s," he recalled, remembering a scene that happened when he was a younger man.

Now, four decades later, he knew quite well that the moment the miners got within reach, reporters would be falling all over them to get their words. Therefore, Pino wanted them to prepare for the tide of media that would break over them like a tsunami.

"The objective is to teach the miners how to communicate with the journalists—how to reply clearly, to cultivate the natural abilities that each of them has," he explained. "For that, along with the textbook, I plan to teach them through videoconferencing, as soon as we have the necessary technology in place."

The end of the ordeal felt near, and the miners' contact with the outside world was improving steadily. Daily newspapers were already coming to them in the capsules. Certainly not all the news was suitable for them to read, especially the articles about themselves. The psychologist Alberto Iturra explained how the information was given to the press and then revised and selected.

But the news wasn't the only reading material the miners asked for.

"They told us to send them the comic book *Condorito*," Dr. Iturra said. "Many of them are great fans of this character."

He was talking about the most famous and traditional Chilean comic about the Creole culture. There was probably not a house in all the country that had not had an issue of *Condorito* lying on the table at one time or another, and the men in the San José mine didn't want to be the exception.

Another request—almost a demand—was for puzzles and crosswords. The tough part was fitting these into the not-quite-six-inch capsule that went down every day, but the people topside managed.

More and more, the incessant trips of the capsules began to include some vestiges of everyday life. But now even the most mundane task was filled with profound significance. Dirty clothes were sent aboveground to be washed by family members. It was a special symbol, a way of saying "I love you." And every family received, along with their loved ones' packets of dirty clothes, a sentiment that made them feel closer.

Pedro Cortez explained that his son, also named Pedro, sent his towels, some shorts, and a polo shirt they had given him. He said that the family was eager to help their son in any way they could, and right now the most useful, most loving thing they could do for him was to send him his packet of clean clothes.

Amid this more relaxed atmosphere, the authorities announced that they were ending the practice of screening and censoring the families' letters to the miners—a practice the family members had not even known about. True, some had suspected, but they had not made a formal complaint. This announcement was made in the tent provided for writing messages, where the family members were advised by psychologists, who recommended the tone and the words to use in their notes.

However, weeks later, it came to light that the screening had in fact never stopped. The authorities had regretted their decision and reversed themselves: "We cannot expose them to receiving bad news," Dr. Jorge Díaz argued in defense of the policy. "Some wives never missed an opportunity to list their problems in intimate detail: telling their husbands about all the household bills they needed to pay, or insisting on knowing when they were going to receive their wages." These, the proponents of screening argued, were problems that the miners could do nothing about—they would only make them worry and raise their stress levels.

THEY SPEAK BUT SAY NOTHING

After long being silent, Alejandro Bohn, the general manager of San Esteban Mining, the company that owned the San José site, seemed to have found his voice once more.

He decided to talk from his office in Santiago. Clearly nervous, he and his partner faced a press bubbling over with questions: What were the working conditions in the mine? Had it been safe for the workers? Many days before the accident, hadn't the miners warned them that "the mine was speaking," as they said in their jargon? Was it true that in June that same year there was another collapse in the mine, in which a worker lost his leg? Would the mine continue operating?

Bohn, amid lights and flashes, revealed that the company would not continue operating the San José mine, and repeated that there had been no prior event that could enable them to anticipate such an accident as occurred on August 5.

"If I had had the slightest warning that a catastrophe like the one we experienced [could happen] . . . I never would have permitted anyone to enter the mine," Bohn tried to explain. "And although it's a little premature, at the moment we have no intention of continuing with the San José site."

"And what will happen to those men and their families while they are stuck underground?" a reporter asked.

"The company is paying their wages, let there be no doubt of that. We have fulfilled our duty through the month of August."

But when asked about pay during the time required for the rescue operation—which, according to the specialists, could be another three or four months—Bohn said he was "in discussions with the authorities to determine which steps to take next."

"Which steps to take?"

There was no further response. Alejandro Bohn was grateful for the media's presence in the company's Santiago office. He said that the company was providing all the information possible to the government, including the background of each of the trapped miners, and that they would continue working with the authorities to end the situation as quickly as possible.

He had not said anything about the miners' families or the company's relationship with them. He knew that the families would not sit around on their hands but would do everything possible to get compensation for their trapped loved ones.

Alejandro Bohn walked back across the threshold of his office, and the press left, dissatisfied with the meager information they had gathered. Alejandro Bohn knew that what happened next could spell the end of his career.

CHAPTER EIGHT

"UNDERSTOOD, DOCTOR"

Minister of Health Jaime Mañalich had become the official spokesperson on the health of the miners. Two days ago, he had begun giving the families and the press a daily report on the physical and mental state of the Atacama thirty-three.

Pleasantly astonished by what he had seen these past days, he said that the miners' health was "extraordinarily good," and had recommended that they walk around and move their joints.

In his daily conversation with the workers, Mañalich told them that for reasons of psychological health, they needed to establish shifts just as in their mining work: a twelve-hour day shift and another at night, and that they do actual work. They needed to get around, move their bodies, which would help them avoid getting blood clots.

"Understood, Doctor," the miners replied from below.

Typically, it was Luis Urzúa, as foreman of the group, who received the instructions. The authorities up above knew that ultimately, he was the one who must pull everyone in the group together and keep harmony until the day they could be rescued.

Along with Mañalich, many other specialists were at the San José site, helping in the aid effort. Among them was an expert nutritionist from Santiago who guided the stages in the miners' diet during this second period: the rescue itself. Now their nutrition was based on solid foods rather than the initial dietary supplements.

"This stage will focus on several goals, which are to establish strict health conditions to avoid infections or illness, to provide them with sophisticated nutrition but with little bulk, as well as to maintain an atmosphere in which the miners can get enough oxygen," the minister explained to the families. "Everything is complemented by gradual and progressive psychological support, which is being provided as we establish more fluid communication with the miners."

Mañalich knew that this step was more difficult yet, because anxiety plagued those on the surface as well as those below. Therefore, he asked for patience and teamwork.

The health minister was getting used to the routine of meeting with the reporters to give them a medical report on the trapped workers' state of health, and he felt comfortable in his new capacity. He said that right now the outlook was rosy, that none of the men had diarrhea, and—without giving a name—that "the person who had a respiratory problem is doing fine."

"They are in high spirits," he said. "We have started off on the right foot on this phase of the rescue."

By all indications, a mission of great importance was being launched. According to the physicians in charge of the rescue, keeping so many people healthy, well nourished, and psychologically balanced while underground was an undertaking without precedent in the annals of medicine.

For the moment, the workers were content and conscious that the rescue was going to take a while, although no one had yet told them how long.

"Yesterday we informed them that they won't be rescued before September but that they can be with their families before Christmas. They have accepted this and are okay with it."

Following the minister of health's instructions down below was Yonni Barrios, the designated medical monitor. Yonni performed the physical exams on each of his companions.

"He's doing a medical evaluation, which will be communicated directly on video," said Mañalich. Barrios, who knew only basic first aid, was the minister's eyes, ears, and hands down in the refuge.

At this point it was important to monitor the overall state of health of each of the miners, with the main concern being to avoid infections and injuries. And through it all, Yonni Barrios understood his crucial role: to watch over the health of all his companions.

SIMILAR EVENTS IN MINING HISTORY

The survival of the thirty-three Chilean miners would be an exceptional accomplishment anywhere in the world, since normally there was little hope of staying alive for more than a very few days after an accident of this nature. But this rescue, meticulously executed by highly experienced people, with the most advanced technology available in the world and with an extraordinarily cohesive operational mechanism, had a good possibility of ending well. There had been precedents.

In the United States in 2002, nine miners spent seventy-eight hours trapped in a Pennsylvania coal mine, almost 240 feet underground. Satellites were used to locate them, and they were rescued alive and well.

Four years later in Poland, a miner was trapped for five days after a cave-in at the Halemba mine, in the southern Silesia region. He survived without water or food and breathed through a crack. He was rescued on February 27, 2006.

On May 9 of that same year, in Australia, after spending two weeks buried almost 3,280 feet down, two miners escaped alive from a gold mine in Beaconsfield. The first five days, before receiving food, they could only drink the water trickling down the walls of the mine. A third man died.

Two years later, in Shanxi Province, northern China, on August 5, 2008, rescue teams managed to bring out alive eight miners who had been buried five days. The workers had drunk their urine to survive.

Two years later, also in China, on April 5, 2010, 115 miners were rescued after being trapped for eight days in a flooded mine in Wanjialing. To survive, they drank dirty water and ate bits of the pine timbers shoring up the tunnels.

With all these precedents, the Chilean authorities knew there was a possibility that these miners could survive—perhaps not all thirty-three, but a large number of them, at any rate.

THEY WANT BEER AND HOT DOGS

"We, the thirty-three who are down here at the bottom of the mine, under a sea of rock, are waiting while all Chile works to get us out of this hell," Luis Urzúa announced with a weathered miner's gritty resolve. Several days had passed since the group's discovery, and the workers were getting restless, which was certainly understandable.

Urzúa was speaking over an intercom with President Sebastián Piñera, who was in his office at the presidential palace. Urzúa took the opportunity to recount what had happened the afternoon of the collapse. "That day was scary," he said. "We felt as if the whole mountain was coming down on us."

Piñera listened, smiling even though deeply concerned over getting the men out quickly.

"What saved us was that the work crew was late going up to the surface for lunch. If it hadn't been for that, the pickup Franklin Lobos was driving could have been smashed," Urzúa said, feeling relief from being able to speak about it. "Then came the dustbowl," he continued, referring to the persistent dust cloud that filled the air and took more than five hours to dissipate. "We couldn't see a bit of what had happened, or the situation we were in, until the dirt cleared. Then we could see that we were trapped by an enormous rock in the tunnel."

The capsules that brought the miners' food also brought them news from the world above. And after reporting to the president on their status and how the accident had happened, Luis Urzúa now took a moment to offer his condolences on the death of Piñera's father-in-law, Eduardo Morel, a few days ago: "Mr. President, we know that you, too, are having a difficult time—not as tough as ours, to be sure, but you have suffered a sad loss." This small gesture from himself and his comrades conveyed their feeling of humility, empathy, and solidarity at the news of the president's misfortune.

All things considered, the miners were in good spirits. This could be seen in the little requests they made to Sebastián Piñera, such as for a bottle of wine to celebrate the bicentennial.

Others asked for beers, *completos* (the Chilean version of a hot dog), and cigarettes. This request brought a laugh from a president, who, though exhausted and overburdened, was ever ready to respond to whatever requests he could, whether from the miners or from their families. But diet was not something the authorities were going to take lightly, what with the alarming weight loss the men were experiencing—for some had lost over twenty pounds.

PROTECTING THE HEALTH OF THE MINERS

Even though the miners seemed be in relatively good heath, the medical team was able to get a more detailed picture from looking at some of the video images taken. They could see that some of the miners were suffering from urinary and dermatological issues.

The medical team were was very concerned about the emaciated appearance of some of the men after going for so long with hardly any food. They were also worried about the men's environment, particularly the dangerously high levels of humidity and the lack of natural light and well-oxygenated air.

Many precautions would be taken to protect the men's' health once it was time to bring them to the surface. The doctors believed that the miners could suffer adverse reactions from the sudden exposure to sun light, infections from exposure to germs, and post-traumatic stress. After having lived underground for months in a very dark, stiflingly hot, humid environment, suddenly rejoining life above ground could be quite a shock to the system.

Health Minister Mañalich wanted to be sure to protect the men's vision from the sudden exposure to sunlight, which could cause serious retinal damage. He was also very concerned about post-traumatic stress disorder, which could affect the men for weeks or even months after the rescue.

The men would first glimpse the world aboveground through special sunglasses that would protect their eyes from ultraviolet rays, and give them a chance to gradually adapt to the bombardment of light. The glasses would effectively block out 100 percent of the ultraviolet rays.

Once on the surface, the men would first be examined in a very dimly lit room to protect their vision. If their eyes grew dry, since they were no longer in the extremely humid conditions of the mine, they would be administered special eyedrops. Their bodies would be checked for any bruises or discolorations caused by their long confinement in high temperatures. But the most serious and lasting health problems could emerge once they returned home and were responsible for monitoring their own health. Skincare and a healthy diet would be most important.

GREETINGS FROM EVO MORALES

Among the trapped was a Bolivian national, Carlos Mamani. The Chilean government made special efforts to help him feel like part of the group.

"Carlos," said President Piñera, "your relatives came from your country and are now at Camp Hope, and I'm telling you now that the president of Bolivia himself, Evo Morales, sends you his support.

The conversation had lasted eighteen minutes so far, and there was still much more to say. Mamani felt reassured by the news, and the knowledge that his own president supported him made him feel as if he were being accompanied spiritually as well.

That night, like a great sigh of relief, the first letters of the thirty-three were released to the families. It was as if a thick cloud of sorrow had lifted from Camp Hope.

The family members broke apart from their little groups and hurried off to a quiet place with these, their most cherished treasures. After secluding themselves, they read with rapt concentration the little notes of love, hope, and relief from their trapped loved ones.

"I miss being able to see you. I miss it terribly, you don't know how my soul suffers being down here and not being able to tell you I'm okay," a tearful Edison Peña wrote to his father.

"[Tell] blondie that I'm happy the pregnancy is going well. Since day one I've had faith that I'll get out—not right away, but I will make it to the birth of my daughter," read the message from Ariel Ticona.

"When I get out of here, God permitting, we'll go and buy the wedding dress and get married in the church," Esteban Rojas promised his sweetheart.

"I am a miracle of our Creator," declared Juan Illanes.

Meanwhile, down in the mine, the men had begun to drink concentrated broth—chocolate-raspberry flavor—and hydrating beverages.

SYMPTOMS OF DEPRESSION

The situation in the mine was worrisome. Conversations between the miners and family members revealed that the confinement and harsh conditions were starting to take a real toll on the men.

"Don't focus on me—I don't want to appear in anything. I just want to get out of this shit-hole," one of the workers said while being videotaped. "I'm tired and exhausted, and I want to be with my family and get out of this darkness that's killing me. I don't even know if it's day or night."

In this and other ways, the miners began to manifest their longings, apathy, and depression. Five of them said specifically that they did not want to appear in the videos that were going to the surface to reassure their families—all they wanted was to get out. This worried the authorities. The men had already survived seventeen days of not knowing what would happen with their lives; this was no time for the doctors to drop their guard.

"'Depression' is the correct word," said Minister of Health Mañalich, confirming the situation with the families after taking stock of the miners' health. He wanted to put them at ease. "The most affected will receive psychiatric treatment, don't worry. We are preparing medications for them," the minister added. "It would be naive to think they can keep up indefinitely the tremendous spirit that they have shown for so long."

The doctors' concerns were indeed spot-on—five of the miners had fallen into depression. They looked a little downcast and at times emotionally discouraged. After discussing it, the medical team decided that antidepressant medications might be in order.

To act with the speed required in such cases, Mañalich and his team prepared a detailed psychological survey and conducted psychiatric interviews for the five who were the most isolated from their companions.

"They've been so strong. . . . We have to talk with them so they don't get depressed," said one of the miners' children.

LAWSUIT AGAINST THE MINE OWNERS

Days passed, and San Esteban Mining continued to answer the questions of the government but not those of the families. "There are no answers," shouted one relative after contacting the company. And so the families organized and announced that they were suing the owners of the mine and the officials of Sernageomin, and had retained a group of attorneys to represent them.

With this legal action, the families sought to have the owners of the mine and the officials of Sernageomin found culpable in the collapse. This would allow the families to go after the personal assets of everyone found legally responsible. From the standpoint of civil law, even the bankruptcy of the San José mine would not stop the miners from getting compensated for their losses and expenses.

The mine owners, for their part, again denied responsibility for the accident. "This is not the time to assign blame or seek pardon," said a harried-looking Alejandro Bohn. "So what *is* this the time for?" was the families' response.

Despite all the prior incidents that began to emerge about the unsafe conditions in the mine, both Bohn and his partner, Marcelo Kemeny, insisted that "everything had functioned as expected." Nevertheless, there were thirty-three lives at risk in the bottom of the mine, and they had survived thanks to the support of their families and the actions of the government—not due to any action of the owners.

As for Sernageomin, some members of the Chilean parliament who were close to the Concertación (the center-left coalition that had led the country for the past two decades and was now the opposition) had appealed not to "demonize" Sernageomin. After all, they pointed out, that agency worked with scarce human and financial resources in many sectors of the country. Others, however, believed that the investigation should be carried through to the bitter end, to establish exactly who was responsible for what.

The lawyer representing the mining company, Hernán Tuane, said that his clients had been made the object of "slanderous accusations." He further announced that honoring such requests as payment of the workers' wages and the rescue expenses could force the company to declare bankruptcy—a statement that many members of the public took as a threat. "The company has a positive balance sheet," he added, "but it has not generated revenue since the collapse."

These statements scandalized politicians of every stripe, and Chilean society's sense of outrage grew, fueled by the statements of the key authorities in the administration. Interior Minister Rodrigo Hinzpeter described the company as "having a lot of nerve," while the administration's spokeswoman, Ena Von Baer, stated that it was "unpresentable."

CHAPTER NINE

THE DIRECTOR

"You can see that things around here have changed, too," said miner Mario Sepúlveda. "We're clean-shaven, and we have new clothes and shoes they've sent us."

Indeed, Mario looked downright dapper in front of the video camera. He was no longer the same bearded, ragged castaway moving phantasmagorically in those first images sent from the bottom of the mine shaft. In this second video recording, Mario had a smooth face and sported a red t-shirt, and his sentences were precise, for precision was just the thing needed at this point:

"This is an excellent team of professionals down here. And that is one of the things that give us the most strength. Here, my friends—and this is a message for all the people," he added in a style that sounded almost political, "the family of miners is not the one we knew a hundred or a hundred fifty years ago. The miner of today, my friends," he said in a clear voice without a hint of hesitation, "is educated. He is a miner you can talk to, a miner who can stick his chest out, who can sit down at any table in Chile."

Mario knew that he was standing in full view of the entire world. The images from the mine were eagerly awaited—more than any prime-time television program. Every time they showed the thirty-three miners, the channels made the most of it, because they knew that the world was watching anxiously. And in these images so full of strength and hope, one face kept showing up, leading the way: Mario, who always seemed to overflow with energy. He did not seem to care in the least that he was nearly half a mile underground. And that breezy nonchalance showed in his attitude before the camera, his impeccable speech, and his assertive linking of the ideas he wished to convey, even when leading a tour through the shadowy place where he had been living for nearly a month. He was marvelous.

The rhythms of the Dominican musician Juan Luis Guerra formed the soundtrack to the video. "El costo de la vida" ("The Cost of Living") could be heard in the background while Mario presented his companions. And the sign-off—which no one wanted, least of all Mario Sepúlveda—came with the ballad "*Burbujas de amor*"("Bubbles of Love"), even as his companions were insisting, "Hey, wait—we *all* want to talk."

But almost all the miners did speak to their families through the video. The only one who declined, without giving reasons, was Ariel Ticona, a shy, reserved man.

Raúl Bustos greeted his wife and two daughters. "We're all doing fine," he said. "We're better fed, and we've been given thermal socks, so as you can see, we are better cared for."

Yonni Barrios, who had taken on the role of paramedic, took the jokes of his mates in stride. "Dr. House," they called him, and one asked, "Are they going to give him a medical degree when he gets out?" The introverted Yonni looked up and managed only a smile as he applied a nicotine patch to one of his colleagues.

The ex-soccer player Franklin Lobos, one of the eldest, on seeing the photo of his family, lost his calm for a moment and grew emotional on camera.

"Here I have my daughters, my two hearts; I have my two lives that make me keep on struggling to get out of here."

The video affected the families deeply. As they watched their loved ones after 35 days in a subterranean chamber, many couldn't keep back the tears. Others were surprised at the good humor and the faith that came through in the images, especially seeing the close of the transmission, when Mario Sepúlveda signed off with another of his messages:

"If, in one way or another, we felt proud of Chilean copper and the Chilean mining industry before, with what it is doing now we are going to feel much prouder. The truth is, dear organizers, dear rescuers, Mr. President, and all the people who make this beautiful [thing] possible, we are very grateful, and today we have given you a demonstration of the progress and of our gratitude. Now the boys must sign off, and I, too, sign off, my beautiful family. I love you very much. I am stronger than ever, and we, the thirty-three, will get out of here, hand in hand—you can count on it."

His words, grand and heartfelt, sent an emotional shiver through the millions who watched in hope.

SETBACK

But it wasn't all good news. For uncertainty always plays a part in rescue operations, especially when considering the geology of the Atacama.

The families asked Golborne, "Minister, what happened? The machines aren't running. We don't hear them working. Why have they stopped?"

The final borehole of the Raise Borer Strata 950 rig suffered its first setback with the detection of a fault in the mine walls in the first sixty-five feet of drilling, forcing it to stop working for several hours. André Sougarret, the engineer in charge of the rescue, saw the problem as easily remedied and termed it "to be expected," but the families were tired and in no mood for setbacks.

Sougarret explained the situation. "From here down to a hundred meters, we expect to see faults, which we will work on as we get to them." The families' impassioned demands came fast and loud, and he mollified loved ones for the time being by explaining that he was ordering the reinforcement of the walls of the shaft through which the miners would come to the surface.

That same September morning, a group of experts from the North American Space Administration (NASA) arrived in Copiapó. They had accepted a request by the government to advise the medical team in care of the miners' health. The delegation included psychologist and expert on mood disorders Albert William Holland, space physician James David Polck, engineer and submarine rescue expert Clint Cragg, and the medical chief attached to the Johnson Space Center in Houston, James Michael Duncan. Once they were fully abreast of the situation, the NASA team could not believe how these men could still be alive and in such good spirits given the situation. As all the world knew, NASA had acquired vast experience on similar isolated situations in its decades of investigation, experimentation, and projects of immense complexity, and the experts were amazed.

"The first thing we'll do," said Albert William Holland, "is create artificial conditions down in the mine to replicate day and night."

With equipment, generators, and LED lamps, his team went to work with the miners to rebuild the previous facilities. "We have daylight!" the thirty-three crowed with joy. "Even though right now it's night outside!" They saw in NASA yet another sign that things topside were functioning, and thanked them in every video-conference, every letter, every telephone call to the surface.

While all eyes were on the team of NASA specialists who had worked so well with their Chilean partners, in Santiago the mine owners submitted a document to the court and called a meeting of the creditors to review the company's financial viability.

The action was immediately interpreted as a covert request for bankruptcy, and the company received a steely repudiation by the government.

"The courts will be the ones to determine the conditions of this bankruptcy," Minister Golborne announced with well-directed anger. "We are also concerned for all the mine's workers' job stability."

ROSARIES FROM POPE BENEDICT XVI

The days went by, and when a month had passed since the ordeal began, the Catholic Church began to take on an even closer and more central role. The archbishop of Santiago, Cardinal Francisco Javier Errázuriz, came to the desert with thirty-three rosaries blessed by Pope Benedict XVI and celebrated an emotional mass with the families and friends of the trapped workers.

"Beloved brother miners, it is with great pleasure that I bring to each of you a rosary which the pope has sent you, after having blessed it personally," the cardinal said. "He accompanies you with his prayers, asking our Father in heaven for the intercession of Our Lady of Candlemas, and of St. Lawrence, who continues giving you strength and hope that the Holy Spirit awakens within you; and that you be rescued as soon as possible. Also, he blesses your beloved families, your companions, and those who work day and night so that you can come out from the depths of the mine. I have also brought you an image of Our Lady of the Rosary with the baby Jesus, our Lord, who are with you and cheer you on. For my part, I commend you from the bottom of my heart. The solidarity and good cheer that you have shown, and the spirit of faith that you embody, have moved all of us Chileans, and you have become one of the greatest gifts for our country to receive as it begins the celebration of its bicentennial."

Visibly moved by the response of the families, Errázuriz, dressed in a black cassock that was already gray with the dust of his surroundings, spoke with the hundreds of people who now filled Camp Hope.

"It is amazing how this event has united us all as a family, how there is no one in all Chile who is not watching in suspense, day by day, what happens to them."

The cardinal continued to stress the workers' great courage and the strength, optimism, solidarity, and discipline with which they had faced this situation. "That which occurred here is one of the greatest gifts of how the Chile of the future must be—the cooperation with strength of spirit, hope, and happiness," he said.

After the mass and after speaking with the parishioners and family members, the cardinal turned to the government authorities and asked to speak with the miners by telephone. "I wish to deliver directly the message of faith and hope that I bring from the Vatican," he said.

No one hesitated a moment, and Errázuriz was put in contact with the miners. The prelate got on the telephone created by Pedro Gallo. Down below, the miners expressed their gratitude, saying, "Faith moves mountains, so how can it not move this one and return us to the world?" They were deeply grateful, and this time there was no laughter and merrymaking below. The emotions were turned inward, toward reflection and silent acts of spirit and prayer.

FIBER OPTICS TO THE HEART

"We can't go on together, with suspicious mi-i-inds," sang Edison Peña. On the black-and-white screen half a mile above him, his wife, Angélica Álvarez, watched him get up and mime some steps as if he had a microphone in his hands. He actually seemed to think he *was* the King of Rock 'n' Roll. Feeling a mix of amazement and happiness, she wept. She knew how much Elvis meant in the life of her music-fanatic husband—if Edison was dedicating one of his songs to her, it must mean he was okay. "Those words are enough," she murmured upon seeing him perform live.

The families had received recorded video of the miners talking to them before, but this was the first time that a live video feed was arranged. The families would get to talk to their loved ones live, hear their voices, and exchange greetings. This video image came on Saturday, September 4, almost a month after the disaster. It was a special day for the inhabitants of Camp Hope. Hours before, technicians from Bellcom, Micomo, and Codelco had run fiber optics down to the refuge of the thirty-three. During the morning they had done tests, and at two P.M. they were ready to make the longed-for first face-to-face communication with the miners.

The families had been notified, and they arrived punctually at the meeting place, obviously nervous, many of them with new clothes and hairdos. The small children were dressed in their best outfits, as if going for an outing. After crossing the safety barrier, each family—limited to three members per miner—climbed aboard a pickup truck that took them to the mine.

Arriving at the site, they came forward, heads held high, as if a metaphor for their hopes. The procession moved ahead as the longed-for moment drew near. They gazed openmouthed at the special tent set with chairs, video gear, and a microphone.

They were told that this first time, family members would get just a little more than four minutes to talk to each miner, who would appear on the large screens in black and white. The trapped men, for their part, would not see their families but would only hear their voices. At once so much and so little, for everyone.

"It's important not to tire them out or stress them," the authorities told the families. "You should not feel downhearted, because that's not what they need. You need to send them only signals of happiness and peace." And indeed, the families were upbeat, and they restrained their desire to talk for hours with their loved ones, obeying the instructions with earnest discipline.

"Pedro, stay strong!" said his mother, Doris Contreras. Leaving the tent, she was all smiles. "I saw my son clearly, shaved and everything," she said to the press.

"I'm worried about the bills," miner Alex Vega said anxiously to his wife, Jessica Salgado, accompanied by the younger of her two daughters, age six.

"Don't worry, I'll take care of it. We love you!" she replied as her daughter peeked around from behind her skirt and said, "I love you lots, Papito."

Wives, brothers and sisters, mothers, and children followed on the pilgrimage, taking their turn to communicate in person for the first time with their missing loved one. As they were leaving the tent, their emotions were obvious. Most of them were teary-eyed. "It's amazing to see how well they are doing after a month underground," one relative said. "I can't help feeling sad seeing him down there," said another. They had come biting their lips nervously, and left with tears that soon dried in the desert wind.

Afterward, everyone agreed that the men looked better physically than they had, and were in remarkably good spirits, although several women asked their men to shave the beards they had grown.

This pivotal encounter coincided with a visit by four of the Uruguayan rugby players who had survived the airplane accident in the Andes in 1972.

They had come to encourage the family members but also to speak directly with the miners, whose situation reminded them of the seventy-two days, thirty-eight years earlier, when they hung in the balance between death and life, despair and faith, fatigue and strength. "We told them they have to fight to get out of there, for their families and for themselves," said a much-moved José Luis Inciarte as he walked about the site with Ramón Sabella, Gustavo Zerbino, and Pedro Algorta.

Amid the festive atmosphere, Angélica Álvarez, nourished with hope, kept in her heart Edison's last words to her. They had been practically forced from him by the next of his companions, who was waiting his turn to speak. "When I get out of here," he told her, "we're going to Graceland." And she was sure that soon they would realize their dream and visit the mansion of Elvis Presley.

PÍA, HEART OF ALL THE WOMEN

The one living perhaps closest to the despairs and conflicts of the women was Pía Borgna, wife of Miguel Fortt, the engineer who had come up with the idea of multiple boreholes. His unconventional proposal, combined with his supreme confidence, had at first earned him the authorities' distrust. But buoyed up by the constant encouragement of his wife, Miguel ultimately had won them over.

Pía was young and beautiful. Her Italian ancestry showed in a striking silhouette that quickly stood out in an insular country such as Chile. Her movements and vocabulary, too, drew a different profile from the one that predominated in Camp Hope. But with her innate intelligence and humility, she quickly fit in and gained the affection of her peers, to become "just one of the women" in that desert redoubt. Pía knew almost nothing about Miguel's profession and even less about the miners' world, but she found just the right expressions, and the right moments to look into the eyes of her new comrades and be with them in their good moments and bad. Since the moment she arrived, she had been unusually compassionate, a good listener with honest answers, a mediator of conflicts, a counselor, a friend.

A telltale sign of her affection was the support that she, along with her husband, gave day by day in the mine. And that affection grew stronger just when it was most needed: in the days before the trapped men were discovered alive. It was a time when gloomy rumors carried more weight than the real news, and as a result, hopes began to crumble. Still, this giver of immaterial goods was not only a spiritual help; she took it upon herself to travel every day into town, to buy food, maté, medicines, and anything else her new sisters needed.

Pía was the confidante of the wives of the trapped miners. "It's just that there are women who don't have great relations with their husbands, and others who were just starting to before the accident," she said with quiet passion. As she spoke, she hid a justifiable pride in embracing a humanitarian task that she had never imagined for herself.

THE PILLARS OF THE MINING FAMILY

The miners' wives were amazingly resilient. The severity of the desert, and the hard and dangerous work that their men did, had forged within them a stoical strength. Many of them were northerners—sisters, daughters, granddaughters, nieces, and grandnieces of miners. They had known that world from being born into it. Others had come from various cities of Chile and adapted to this new life while keeping their customs, speech, and traditions of old.

The women were self-sacrificing, not only in taking complete care of their children and keeping the household together, but also in being the spiritual and physical support of the men who dug the earth for their sustenance.

They felt daily the ancestral machismo that had imprinted generations and that, although now in a less violent form, still persisted.

They treated their men, their providers, with special care, and food was an important theme, an extension of their love and affection.

In the north of Chile, many fruits and green vegetables were scarce because of their high prices. So these women nourished their families with foods that were abundant and high in energy. Lots of carbohydrates and beans on the table.

In the north it was common to mix meats with potatoes, rice, and fried foods in the same dish, contrary to what nutritional experts advise. But optimal nutrition was not always available, or affordable. *Charquicán* (an almost exclusively Chilean stew of ground meat, potatoes, pumpkin, and green vegetables); *tomaticán* (meat mixed with potatoes, fried onions, and corn); and beans, lentils, peas, and garbanzos—nearly always with rice and eggs—made up much of the daily diet of these northern homes.

Breakfast was something else. A long tradition in Chile insisted that it be rich in calories, perhaps due to its Anglo-Saxon origins.

The man must be well fed, and when he came home, he might be greeted with a beer—ice cold in summer—or else a glass of wine, served the moment the key clicked in the front door.

Part of this tradition showed in the way Camp Hope was organized, through the leadership of María Segovia. Everything had a place and served a function. So strong was their commitment that the women had been able to set aside any differences, quarrels, conflicts over money, or other difficulties. For such trivial things paled before the greater common good: seeing their men alive again, above the ground, and embracing them like never before.

YONNI'S DOUBLE LIFE

Marta Salinas faced an uncomfortable situation. She was the lawful wife, for twenty-eight years now, of Yonni Barrios—but Yonni had been living for months with his new woman, Susana Valenzuela.

Marta was one of the first to show up at the mine, but when Susana's family came, the situation for Marta became simply untenable. She couldn't bear the pressure and now was thinking of returning to Copiapó, where she owned a small grocery store.

Marta Salinas was initially sure that the thirty-three were dead—so sure, in fact, that she stayed on at the mine mainly to see that they recovered and delivered the bodies. But now she knew the miners were alive, and Yonni was perhaps livelier than most.

Meanwhile, Yonni Barrios, never one for marital strife, wanted both Marta and Susana to receive him when he emerged from the refuge.

But to this questionable and puzzling request, his estranged wife's reply was categorical and dead set. "I'm happy that Yonni has been saved; it's a miracle of God. But when he comes out of the mine, I won't be there," she said with long-suffering dignity.

September 8

The stress, worry, and depression were launching a major assault down in the San José mine. Some of the workers had not wanted to use their precious minutes of videoconference with their families, in protest over the shortness of allotted time and the censorship of the messages sent from the surface.

"This is infuriating," Víctor Vargas said to his family. "Everything's a disorganized mess, and there's almost no time for talking to our families. We're in the dark about what's happening up there."

On the surface, meanwhile, many of the relatives were also showing their annoyance over the rigorous censoring of their letters. "Mr. Minister, why aren't our letters getting to our men?" cried the angry wives. But Mañalich didn't lose his composure at these complaints, much less let himself be swayed from the course charted for a safe rescue, though he certainly understood the families' frustration.

Despite the position of the authorities, the families talked among themselves about the distress and anxiety that the miners must surely be feeling over their rescue.

"I want to be the first one out," Carlos Mamani, the Bolivian, had said to his companions. "I want them to rescue me the soonest they possibly can—I can't take any more of this." Verónica Quispe, his wife, was clearly concerned over her husband's words. "He'll be okay," she said. "He's just making the point that he wants them all to get him out of that hole, just the same as some of his Chilean colleagues have been asking." Verónica demanded nothing less than the presence of Evo Morales, the president of Bolivia, at the San José mine, and when he came to Chile for the bicentennial celebration, he visited her on September 18.

"What many of us feared is happening: they're getting distraught," Nélida Villalba, the wife of Pablo Rojas, said to the authorities.

After hearing these complaints from families and workers, the minister of health agreed to lengthen the communication time to five minutes per family. "It's a humane concession—a slight alteration that won't affect the guidelines we're setting," explained a member of the minister's advisory team.

UPSETS AND BOTHERS

When the families' requests for more time had been resolved, the titles rolled in this cinematic event taking place both above and below the driest ground on earth.

Miner Mario Sepúlveda remarked to his father that life together had become a bit difficult in Camp "Los Treinta y Tres," as they had dubbed their little settlement at the bottom of the mine.

"He told us they had been having some problems down there, but that they'd gotten it all sorted out. They held a meeting, in which they agreed to avoid those problems in the future," said Mario's father.

Psychologist Alberto Iturra attributed the troubles to the lack of coordination in the videoconferences. "It was because of the tight scheduling for the time they needed to talk—because they were having to make such short conferences," he explained.

Meanwhile, the National Emergency Office released a brief communiqué reporting that one of the miners suffered from a "stress-related problem" stemming from a previous pathology and that he was being treated with medications. The office also acknowledged that another of the miners had suffered from an intense toothache, for which he was put on antibiotics.

One of the NASA experts, Albert Holland, who had been staying in the camp, spoke of the reigning anxiety and warned of the tough challenge that awaited the miners once they emerged into the outside world. These men were expected to be freed, liberated from the bowels of the earth, but their reintegration with the outside world after the long isolation would not be easy. It was a subject much discussed every day in every part of the country, and everyone had something to say.

"There is a great focus on the immediacy of getting them out of the mine," the American psychologist explained, "and now they are starting to adjust their thinking. Specifically, they're rethinking whether this is a sprint or a marathon, and that a marathon has a very different rhythm; it requires strategies. . . . When the miners get out, they will have a well-earned status as celebrities in their country, and there will be pressures, from society, from the media—all these people wanting part of their time. They and their families will need protection, medically and psychologically, during the first twenty-four to forty-eight hours after their rescue."

THE TECHNOLOGY BEHIND THE RESCUE

Another hero in this story is the Schramm T–130, better known as T–130, or Plan B. That's the machine that drilled for thirty-six days to reach the miners and formed the shaft that would allow the Phoenix rescue capsule to travel down 2,300 feet to reach the men. Once it had accomplished its mission, it was ushered from the staging area to the loud applause and cheers from the hundreds of people gathered at Camp Hope.

The farewell to the drill seemed more like a raucous national holiday celebration than an expression of thanks to a rescue team trying to save lives.

Personnel from Geotec Boyles Brothers, a Latin-American affiliate of the North American Layne Christensen Corporation, were responsible for transporting the T–130 to the San José mine. Two North American drilling experts, Jeff Hart and Matt Staffel, flew to Chile from Afghanistan, where they had been drilling water wells for the U.S. military. They were in charge of operating the drill at the mine.

According to his coworkers, Jeff, who has been operating drills for twenty-four years, is one of the best in the world. He was very affected by the emotional scene at the San José mine. While he was working at the mine he said in a phone interview with a reporter, "It's incredible. It's been an awesome feeling. It's been exciting, an emotional rollercoaster, for sure." He and Matt took turns keeping the drill in continuous operation for twenty-four hours a day. Early estimates projected that the drill would not reach the miners until Christmas, but Jeff was instrumental in getting the drilling completed two months ahead of that estimate.

Describing the scene when the drill made it through, Jeff said, "We had a video we could actually see from the mine, the tool coming through into the mine, we knew we had contact. . . . As you can imagine, it erupted. We have been here busting our butts, we have worked every day, we fought the odds. We fought till the very end, we made it into the gallery, it's incredible." He said it would be a story to tell his grandchildren one day. "This is the most important work I'll ever do in my life," he stated.

While Jeff finished up the drilling, his wife Dora followed all the developments from their home in Arvada, Colorado. "Jeff can do anything he sets his mind to," she said proudly. She kept in contact with her husband through email every day. "He's not looking for the fame of it," she told reporters. "He's just trying to help where he can."

Dora confessed that before making the decision to travel to the desert in Chile, they hadn't known much about the country. "We knew about the accident there, and then we realized that he could help."

The night before the drill broke through, Jeff sent Dora an email letting her know that they were very close to reaching their goal. As soon as she read that message, she got up and turned on the television to watch the news stories. In the middle of the desert, thousands of miles away from home, and after answering questions from dozens of reporters, Jeff was finally able to talk to his wife and tell her how relieved he was to have successfully finished the job.

They would have to be apart for a little while longer, however. Jeff enjoyed the outpouring of kindness and sincere gratitude from the Chilean people, while Dora believed that people around the entire world were praying for the miners and their safe rescue.

Right after the drill broke through, the other hero of this part of the story, Matt Staffel, 29, tried to understand the questions flying at him in rapid Spanish from reporters and family members of the miners. Born in Denver, Colorado, he had also been in Afghanistan drilling wells for the U.S. military before traveling to Chile to take part in the rescue. He had spent four years operating machines similar to the T–130 in Afghanistan. He wasn't sure if he would return there or go back home once he left Chile.

The drilling advanced according to plan. While the Raise Borer Strata 950 had punched through 370 feet of rock, the Schramm T–130—designated plan B—had gotten to 403 feet. However, it would need to make a second pass to widen the hole.

"The T–130 starts out drilling a guide hole, so it has less material to chew up. But then it has to take another pass later to widen the diameter, so that means twice the work," the authorities explained to the families. But the families didn't care about such details; they just wanted to see their men aboveground, alive and safe—and soon.

Meanwhile, the people in charge of the rescue worked determinedly, installing a platform to support the huge, heavy oil-drilling rig—plan C—which would arrive in pieces, carried to the mine on forty-two tractor-trailers.

Minister of Health Jaime Mañalich announced that a hospital tent and a heliport would be set up a few yards from the mouth of the mine, to attend and transport the miners as soon as they were rescued, presumably in mid-October. Everything was falling into place for the final push.

Álex Vega, the tenth of the thirty-three miners, rescued on October 13. His father, José, himself a miner, was one of the first rescuers onsite after the collapse of the mine.

Head engineer André Sougarret, far right, goes over the plans for the rescue with Chile's President Sebastián Piñera, second from right, and other members of the team.

Mass for the family members at Camp Hope with Bishop Gaspar Quintana Jorquera. As they waited, the worried families erected memorials for the trapped men, lit candles, and prayed. On a nearby hill overlooking the mine, the families placed small shrines for the missing men, with flags, pictures, and statues of patron saints.

The mine rescue effort was dubbed *Operación San Lorenzo* after the patron saint of miners. Here, Piñera poses with the rescue team prior to the operation to bring the trapped miners to the surface.

The first
test of the
Phoenix
2 capsule,
which would
bring all
thirty-three
miners
safely to the
surface. It
was painted
the same
colors as the
Chilean flag,
red, white,
and blue.

The first rescuer,
Manuel González,
enters the Phoenix
capsule and descends
into the mine. He
will also be the last
rescuer to emerge
twenty-four hours
later.

Bolivian President Evo Morales expresses
solidarity with Piñera after giving a speech at
Camp Hope on the day of the rescue. Among
the trapped was a Bolivian national, Carlos
Mamani.

Franklin Lobos, the twenty-seventh rescued miner, and a former professional soccer player. He emerged from the mine clutching a soccerball.

The group's spokesperson and morale keeper, Mario Sepúlveda, pictured here with Piñera, was the second miner to be rescued. He came to the surface with a sack of rocks from inside the mine, which he handed out as souvenirs. Dubbed "Super Mario" by the media, he had been the host of the miners' video journals.

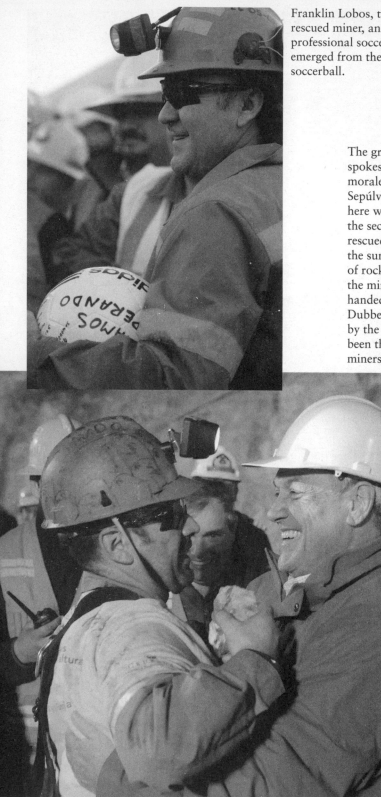

Omar Reygadas, the seventeenth miner to reach the surface. Later he would admit to feeling nostalgic for his bed and the little makeshift desk, where he had written so many letters to his family.

Osmán Araya, the sixth of the rescued miners, told his wife and baby daughter Britany in a video message: "I will fight to the end to be with you."

Esteban Rojas prays immediately after coming to the surface. He told his sweetheart while still underground: "When I get out of here, God permitting, we'll go and buy the wedding dress and get married in the church."

The fifth rescuer, Patricio Sepúlveda, prepares to descend into the mine.

Ariel Ticona was the group's communications specialist, maintaining the telephone equipment that was the miners' link to the outside world. The thirty-second miner to reach the surface, he missed the birth of his daughter, Esperanza (Hope), while trapped underground.

Yonni Barrios, the twenty-first miner to reach the surface, was nicknamed "Dr. House" for his role as the group's medic. Here, he is embraced by his mistress, Susana Valenzuela. Having only just learned of his years-long affair, his wife declined to attend the rescue.

The last miner to come to the surface was Luis Urzúa, the foreman and leader of the group. He told Piñera, "I've delivered to you this shift of workers, as we agreed I would." The president replied, "I gladly receive your shift, because you completed your duty, leaving last like a good captain."

President Piñera visits with the thirty-three miners in the Copiapó hospital, the day after their rescue.

CHAPTER TEN

HOPE IS BORN

"Hurry it up! We'll be late!" María Yáñez yelled to Héctor Ticona. The parents of Ariel, one of the thirty-three, rushed up the steps of the Copiapó clinic. They didn't want to miss the birth of their third grandchild. A few steps behind them, the news photographers followed, cameras at the ready.

The baby came into the world by cesarean section at 12:22 P.M. She weighed eight pounds, two ounces, and measured 18.9 inches long.

"She's just like her papa," the elated grandmother said on seeing her. "I'm so excited, but I feel a little sad that my son couldn't be here for the birth."

Inés Yáñez, the miner's aunt, felt much the same: The news filled her with conflicting emotions. "I would have liked for my nephew to be here," she said. "So we're here to lend support. She looks a lot like him: her little face, her hair . . ."

It was the moment the family had awaited anxiously—all the more so when, after eight months of pregnancy, the mine collapsed, separating Ariel from his wife and his two boys, Steven, five years old, and Jean Pierre, nine.

"My sister-in-law cried all night before the birth," said Victoria, Ariel's sister. She never imagined that on this day he would still be trapped in the mine.

The joyful moment in the clinic was captured by all the television cameras that could fit in the room. One of the reporters got the images to the family, who rushed them to the mine. At five in the afternoon, the video went down in the capsule, and Ariel, finally and for the first time, saw his daughter. With tears of joy, the father got congratulations all around from his companions.

The same day that Ariel Ticona's daughter was born, the rescue experts and some engineers from the Chilean Navy made the first tests of the rescue cage that would be used to hoist the miners to the surface. Miguel Fortt, the engineer who had come up with the plan for multiple boreholes, explained to the families the design of the capsule that was going to bring their loved ones up.

"It's shaped like a frankfurter," he said, drawing in the dirt with a stick.

The cylinder, designed by engineer Alejandro Poblete and fabricated in the Chilean Navy shipyards in the southern city of Talcahuano, had already seen duty in other mine rescues. Many referred to it as the "cylinder of life," because ultimately, that's what it was.

The capsule consisted of a sheet of steel 4 millimeters thick, 21.25 inches in external diameter, with two conical ends. The maximum length of the capsule would be a little over 8 feet 2 inches. The empty weight was estimated at 670 lbs.

It had a special feature of eight retractable wheels on its sides to keep it from rubbing against the sides of the borehole. The idea was to soften the bumps during the journey up the shaft. Every little thing had a purpose in the operation.

Inside the capsule, there was an "on demand" compressed air breathing system, delivered via a direct connection to the airway. The compressed air was supplied from 165-liter bottles, each good for three hours' duration. And if anything should go wrong, the vehicle was equipped with an escape system. Its bottom could be released, and that way the passenger could descend by rope back to the bottom of the shaft.

To stabilize the passenger inside the capsule, there was a harness that had four points of attachment, with quick release, securing the passenger by the shoulders and crotch. The system worked for lowering as well as raising.

"We're going to test it until we're totally satisfied with the results," Fortt said to the families. His improvised diagram, etched in the dust of the camp road, reassured them that the thing would work.

In the same moment, miles away in the clinic, the name of the newest member of the Ticona family came instantaneously. It seemed almost foreordained. Upon learning that it was a girl, the parents had wanted to name her Carolina Elizabeth. But after the accident, those plans changed, and it was Ariel himself, in one of his letters, who suggested another, more symbolic name, one that better expressed what was happening in their lives at the time. The baby's name was Esperanza—Hope.

AN EARLIER RESCUE DATE

A month after the collapse, despite rumors, suppositions, and natural pressure, the rescue teams seemed to be achieving their objective.

At this point, the popular minister of mining, Laurence Golborne, confirmed enthusiastically that the rescue date could be moved up to the middle of October, although he ran the risk of terrible consequences if he couldn't make good on his promise. The big news exceeded the expectations of earlier days, which placed the rescue date sometime in early November.

The rescue workers chosen to bring the thirty-three miners up to the surface prepared for the critical task with simulation exercises. They practiced handling the Phoenix capsule, and created an experience that would mimic as closely as possible what the claustrophobic trip down the shaft would be like.

The team of rescuers set up shop along one side of the San José mine, as far away as they could get from the family members and reporters who had flooded into Camp Hope to cover the story that had captured the attention of the whole world.

Distinct tasks were assigned to different workers according to their professional background. The Navy officers, for example, would go down to check on the physical condition of the miners and would bring a wide variety of medicines along with them. Because of their rigorous training, they were even prepared to give medical attention to heart attack victims. They would bring medications to treat anxiety, only to be used as necessary. The plan was for the first miner to go up in the capsule as soon as the Phoenix touched down in the mine.

All of the rescuers had been selected because of the specific training and experience they each had and their extraordinary ability to remain calm and function in extremely confined spaces. In the last hours before the rescue attempt got underway, the rescuers pored over medical reports and psychological profiles on the miners and studied the various lists of medications that the miners were taking.

The secretary of mining notified the families, assuring them that after overcoming several geological problems, the machines were in a position to achieve their objective very soon after the fifteenth of the month.

Golborne humbly explained that they had made it through several critical zones in the drilling, which led them to be cautiously optimistic about the time needed to make the rescue a reality. The families began to prepare themselves for the return and rebirth of their loved ones.

AN EMOTIONAL BICENTENNIAL CELEBRATION

It was midday, and the harsh sun of the Atacama Desert beat down on the dozens of people gathered at Camp Hope to kick off the celebration of the *Fiestas Patrias*. There were no shady trees or parasols, no sunshades or umbrellas—just a hot day getting hotter as the people gathered.

A police choir, its members in impeccable dress uniform, sang the strains of the national anthem, which the families of the miners followed.

Sweet fatherland, accept the vows,
which Chile swore at your altars:
You'll either be the tomb of the free,
Or the refuge against oppression!

The families, friends, and few journalists and workers congregated there sang out the words with strength and feeling. It was a moving ceremony.

The national flag was raised on a standard with a red base, in memory of the painted drill head of August 22, which had confirmed to the world that the workers were alive after seventeen days of anxious wondering. Meanwhile, throughout the rest of the country, millions of citizens congregated in plazas and restaurants to sing in unison the national anthem, many of them no doubt thinking of their countrymen trapped deep in the earth.

It promised to be a special celebration. For one thing, carnivals and huge festivals were something of a rarity in Chile, unlike the rest of the continent. But today the whole country would turn out to mark its two centuries of independence. It also was a special occasion because the thirty-three men underground had moved the entire nation, unifying it more than ever through the shared feelings of concern, solidarity, strength, and hope.

CREOLE GAMES

In Copiapó, with the traditional rites of the republic out of the way, the city turned into one big party. In the north of Chile, this meant setting up little streetside arcades, with such games as throwing rag balls to knock over "stubborn cats" (milk cans painted with feline faces and filled with rocks so that they didn't fall so easily). Also, there was the top-spinning competition. This Creole game combined skill with cunning, with the objective of keeping your pear-shaped top whirling on the concrete or dirt until it outlasted all the others.

And there was the ever-popular game of ring toss. Standing behind the foul line, the participants honed their marksmanship and tried to pitch the wooden rings over the neck of one in a set of bottles. Whoever succeeded took home the liquor that his ring landed on.

With prizes in hand (or resignation on their faces), the Copiapinos at last heard the music for the first *cuecas*—the Chilean national dance. These traditional dances didn't last long, though, for the dance floors were soon given over to the high-energy *cumbias*, which fired the spirits and occupied many happy patriots for hours. Not until dawn did the dance music finally wind down.

The jubilation up on the surface contrasted dramatically with the festivities of the thirty-three trapped in the San José mine. Though they had no music, no beer, and no laughing crowds, they could at least enjoy one of the country's tastiest culinary rituals: traditional empanadas, a finger food that every northerner had eaten since earliest memory.

Made under the rigorous control of the nutritionists working with the medical team, the empanadas sent down to the refuge were a bit different from the usual rectangular version (triangular if made with chile sauce) now being devoured up on the streets of Copiapó. These were long in shape, with reinforced edges, much like a *chaparrita* (a sausage wrapped in a cheese batter). Still, they were made with the classic ingredients: flour, egg, vegetable oil, meat, raisins, olives, and onions, and for the miners they were a nostalgic taste of home. And for dessert, they had papayas, a fruit abundant in the Chilean region of Coquimbo, a few miles from Copiapó.

NO ALCOHOL—FINAL DECISION

The authorities would not budge on the subject of alcohol. If they allowed the miners to have the drinks they requested, they believed it could generate a serious imbalance in the group dynamic.

The chief of the team of psychologists, Alberto Iturra, explained the reasons: "We are not having a party; let's respect the families. The area where the workers must interact is an uncontrollable environment, and we shouldn't add a new variable to what is already a difficult situation."

Dr. Iturra also pointed out that the country had half a millennium of mining history, and the rule had always been that no alcohol was consumed in the mines, for very good reasons. "Also, the high humidity makes one sweat more quickly, and the levels of alcohol are not metabolized in the same way as in an external environment," he explained, leaving no room for argument. "What would happen if one of the miners didn't want to drink anything and gave his ration to his buddy? Who can control that? It could turn into something complicated, and we have enough risk factors already without adding new ones."

FINAL CONTACT WITH
THE REFUGE

It was Saturday, October 9, and the fog blanket over the Atacama was thick—thick even though dawn was not yet more than a dull, icy gleam in the air. It seemed that the dense, low mist known as *la camanchaca*, rather than take its usual route northward, preferred to stay and witness what was presumably to come.

The past few days, the sun had burned away the mist early, but that day was more humid than usual. Day broke, and dozens of little stoves flared to life, fighting the chill and heating coffeepots and kettles for tea and maté.

The warming cups were passed from hand to hand in the various tents of the camp amid an uneasy, tense silence. No one had gotten much sleep last night, because it was more certain than ever that the T–130 drill rig was about to break through the hard crust and reach the buried men. And this thought alone was proof against the chill. The only thing anyone wanted to do was listen for the loudspeaker that would alert them all to the grand moment. With coats zipped up to the throat and wool caps over their ears, some hopped nervously or ran in place to generate something resembling warmth, while others breathed into their cupped hands to warm them.

"Time to break the vigil, kiddos!" yelled María Segovia, the famous mayoress of Camp Hope. "Everyone who can, come and join in. We're going to pray with a pastor who will stay with us until we hear the drillers hit the bull horn—like they did when they found the *niños* alive!" It proved to be an almost eerie prediction of what would occur only minutes later.

Nearby, Jessica Yáñez, whom the miner Esteban Rojas had proposed to via letter, was trying to hold it together after staying awake all night. At the same time, she was also mentally preparing for whatever might happen next.

"We just need to have a little bit more patience," she said hopefully. She was already thinking about the honeymoon that Esteban wrote about in his latest letter from the mine. "We're going to go far away—as far as we can, where there aren't any reporters," she said with a mischievous chuckle.

So this was the mood in the camp: excited energy tempered with plenty of caution. Then, at 8:05, the long-awaited sound blared from the loudspeakers.

Some people jumped; others shouted. Several held themselves back, doubting in silence before finally letting themselves believe they were not dreaming or imagining. But it was true. The horn blared out. The men's salvation was nearer at hand than ever before. The miracle had arrived.

The T–130 drill, which had been boring away for thirty-three days into the tough rock that held thirty-three men prisoner, had punched through. Finally, it had made contact with the refuge. On breaking through into the corridor with the entire drill head, it had completed the 2,040 feet of vertical shaft through which the workers would soon emerge. In a striking coincidence of numbers, thirty-three days of drilling were over—one day for every man trapped underground.

CHAPTER ELEVEN

THE SOUL'S BEAUTY IN THE FLESH

With the rescue of the miners seemingly imminent, the women of Camp Hope began a frantic race to change their appearance.

They wanted to be beautiful when their husbands returned to them.

"I hadn't done anything with my hair," said the wife of Claudio Yáñez. "Now I put in highlights, got a haircut . . . I'm waiting for him, nervous as hell." She smiled at the same time that her eyes were filling with tears.

It was just that after two months without seeing Claudio, she wanted to be pretty for the man she had shared her life with for the past ten years, and with whom she had two children: one eight years old and the other a year old.

But the women had different worries now. They asked themselves how their men would be when they returned, and they were afraid of the answers. They knew that they had to start over, making a new courtship, a new relationship. Everyone knew that their husbands would not come back the same as when they had left for work that fateful August day.

"They will have to adapt to us, not us to them," said one. Another said, "We have to take more care with them when they come back home; we have to keep them from getting too upset." "Damn, my husband's going to want a drink, and I don't want him to, because he's going to be so weak," said a third. They each had their worries, with no one knowing what it would be like to have their men back home again.

How to manage this long-awaited encounter? A thousand questions ran through their heads as each tried to find the way to face this new event, this new stage in their life as a couple. But like the decisive, empowered women that they were, they went to work confronting their fears and chasing away their phantoms.

Some prayed more devoutly than ever before in their lives, while others went off to buy lingerie—a bit of play and relief from all the nervousness. Some came back from Copiapó with garter belts, others with chemises. So powerful and important was this theme of intimate reunion that sellers of lingerie even came to Camp Hope. They sold out in no time. And so it was that the wives, while preparing for the reunion with their husbands, also got reacquainted with their own femininity and coquettishness.

THE ADOPTED FOREIGNERS

The wife of the Bolivian miner, Carlos Mamani, had painted her nails maroon and cut the long black locks that were the envy of hundreds of women among the caregivers, family members, and journalists who congregated at the camp.

"I want to look good," said Verónica Quispe, twenty-five years old, married for the past five years to Carlos Mamani, who had come to work in the mine only five days before the collapse. "Even with all the suffering we have been through, we feel that this has changed our lives," she said with conviction. "Maybe God has done this to improve our lives." Like the other wives in the camp, she, too, was anxious about seeing her husband again.

Even though she came from a place geographically close, Verónica was now in a foreign country, which had had its differences with Bolivia for centuries. Because of this, she had initially felt another fear that the others did not: that of being segregated. But none of that mattered to anyone, because in this eternal vigil, the best of human qualities had prevailed.

Verónica, with her stories about Carlos, her recollections of her life in Bolivia, her hopes and plans for the future, had found such a warm reception that in no time she was a full member of the "clan" of Chilean women.

THE STRUGGLE FOR THE COMMON GOOD

These women—mothers, wives, daughters, partners, girlfriends, live-in lovers, Chilean and foreign—now had intertwining lives, thanks to a change that began on August 5, 2010, when the roof of the San José mine fell in.

Then came the moment when the differences, competitions, and petty feminine rivalries disappeared. Maybe not forever, but long enough to struggle for the common goal that united them: to see their men alive once again. Their desperation, anger, and sadness were thoroughly human feelings that could be soothed somewhat by nightly prayers and the faith that things would turn out well despite the dire early predictions of the experts. And the temple of that unity and mutual support, where they did their best to carry on their usual household routine, even with their kids along for the ride, had been the famous Camp Hope, filled with that feminine stoicism that humanity cannot go long without.

CHAPTER TWELVE

"WE'RE READY, MATE"

As soon as he heard the long wail of the siren, Raúl Valencia, the teacher at the little school established at the mine, jumped out of his sleeping bag and ran barefoot to the bell used to call his few students to class.

He rang it for all he was worth, for almost fifteen minutes. "My ears rang for a while after that," he recalled, laughing at his spontaneous way of proclaiming the good news that morning.

A few yards from the schoolmaster's noisy ringing, Wilson Ávalos, brother of miners Florencio and Renán, watched everything as if in slow motion, while tears traced little salty streams down his cheeks.

He was one of the first to learn of the happy outcome, when a friend who worked on the rescue effort called him. "We're ready, mate," was all he could hear on the cell phone, but he didn't need to know anything more.

"All the sadness is finally coming to an end—or maybe you think I'm going to miss sitting under this tent flap all day," he said, his face lighting up with a grin.

The bread dough and goat cheese for breakfast sat forgotten on the table. All around, emotion seemed to give everything a deeper, brighter hue. Family members, friends, and reporters hugged as if sealing an eternal pact, for in this moment they were no longer relatives and media professionals, but human beings who had been waiting a long time for the same thing.

Farther off, on the other side of the safety barrier that marked off the zone of the rescue efforts, the American driller Jeff Hart was being congratulated by his workmates. His hands had been on the controls guiding the final shift of the T–130 that had broken through the rock and made contact with the miners.

"It was an overwhelming feeling," said the forty-year-old native of Denver, Colorado. He had been drilling for water in Afghanistan when his company, Geotec, the owner of the T–130, brought him to Chile to drill in search of life.

At his side, geologist Felipe Matthews, with the joy of someone celebrating a wedding or ringing in the New Year, popped the cork on a bottle of champagne and spewed its fizz over all the engineers and drillers. The ground, too, lapped up the celebratory foam.

Happiness drenched everyone as a senator from Atacama picked up the cork, now like no other—a historic symbol that must be saved. Any little thing would serve in the future as a memento to capture and record that powerful moment.

The celebration lasted only minutes. The momentary acknowledgment of triumph was over, and immediately began the most crucial part of the work: finishing the job and bringing home the thirty-three.

Hours later, at three in the afternoon, while the heat bounced off the polished rock surrounding the encampment, a sound alerted the inhabitants. The barrier to the work zone was lifted, and an ambulance roared in with sirens blaring.

The emergency vehicle, escorted by two motorcycle policemen who kept looking back over their shoulders, blazed a path while another pair of policemen brought up the rear. The unexpected caravan sped across the only street in the encampment as an enormous dust cloud rose behind it.

Everyone hurried out to look, hearts skipping. Could this mean there had been an accident in the rescue efforts? Would there be injuries? Would this hold everything up? These were a few of the fears that whirled amid this new confusion.

But to everyone's relief, the doubts dissipated in a few minutes. It was all a drill. The objective was to measure how long it took for an ambulance to get between the encampment and the hospital at Copiapó. The authorities were just planning ahead in case, on the decisive night, the thick gray blanket of the *camanchaca* should prove stubborn in retreating. That would make helicopter flights dicey, and nothing could be left to chance.

A little later, the unmistakable racket of a helicopter agitated the family members all over again. It was the first test run for the nocturnal flights—they were running drills for all possible conditions of transport that they might have to deal with on D-day.

The journey would start at the heliport built on a nearby hill and would end at the military installation in Copiapó, since the hospital lacked a place for the copters to land. From the military base, the miners would be moved on gurneys to the health center. The preparations were calculated to near perfection.

"The pilots are practicing their night flights, coming and going, to get an exact fix on the landing zone," Minister of Health Jaime Mañalich emphasized. "This way transport of the miners won't be interrupted while the rescue is going on."

Regarding the worrisome *camanchaca,* which would put the ambulances on green alert, Mañalich made it clear that ground transportation would be only a last resort due to the risk of carrying the recently freed workers over the winding, curvy, rock-strewn road to Copiapó.

"In case of the *camanchaca,*" he said, "the miners will remain in the tent hospital that has been set up at the mine, where eight at a time can be treated. Then, when the sky clears, we can begin to transport them."

They worked at full speed ironing out details at the hospital in Copiapó. The entire private wing, on the second floor, was prepared to receive the patients. They also readied the third floor with seventeen beds and a team of specialists including internists, a dentist, a psychologist, and an ophthalmologist.

Another link in this perfect chain of rescue was security. More than forty police would safeguard the hospital, and traffic through the emergency entrance was restricted. No one but family members, identified by a colored bracelet, could approach the miners.

Everyone knew, though, that at the moment of truth, these measures would be upended by the stampede of the press, right up to the very beds of the thirty-three. This was a foregone conclusion, but it was important to try to keep the rush from getting out of hand.

Along with the predictions, uncertainties, and bets going around after the breakthrough, another announcement was on the way, it seemed. Minister Golborne was at the press area, standing before dozens of television and newspaper cameras, recorders, and microphones. The authority who had been staying here right alongside the families every day since the collapse had something to say. The passing minutes seemed an eternity, and everyone fell silent, trying to read the message in Golborne's face. It was as if all the entire noisy, boisterous press corps had frozen.

"After reviewing the inspection videos and taking geological tests, we have arrived at the conclusion . . ." His words seemed to pause in the air amid the desperation of all to hear the conclusion of his message: ". . . that it is not necessary to sheathe the entire rescue duct, as we had considered doing at several points."

Shouts, cheers, whistles, and more shouts—no one listened to anyone, and everyone to everyone, as all voices rose in jubilation.

The minister's announcement was huge, because now the rescue would go much faster. The fears of another week's delay in the work were at an end. Instead, the favorable scenario would permit them to install the casing—the metal tubes protecting the carriage on its repeated journeys up and down the shaft—only in the first 315 feet. That was the most dangerous section because it was nearest the surface and the most fractured and at risk of rockfall.

"There are no cracks or weaknesses farther down," Golborne explained. "It is clearly in the uppermost section where a small rock could possibly be dislodged. There is nothing complicated to manage, but the process of encasing those first hundred meters will protect against any contingency." He seemed markedly calmer than before, and his breathing was a little more relaxed. Everything seemed to be coming together to speed the freeing of the thirty-three.

WHO'S FIRST?

From the very instant the T–130 finished drilling the most important borehole in its existence, the authorities and the rescue bosses discussed an oddly important subject: the order of the miners' departure to the surface.

The doctors held a clear position: first up should be the most physically fit, the strongest, so that they could test the hoisting mechanism. "We cannot run the risk of one of the more weakened ones getting trapped for minutes in the tunnel," explained Dr. Jorge Díaz, chief of medical operations.

But the people from the government had other ideas. They wanted the second miner up to be Mario Sepúlveda, the "host" of the videos inside the mine, and a face known to everyone, since this subterranean dramatist had been the pillar of strength supporting the spirits of his companions.

The representatives of the administration did not want to waste the understandable opportunity for President Sebastián Piñera to shake the hand of the emblematic Sepúlveda at the start of the rescue, before the cameras of all the foreign television channels, which had been closely following all the details. It had to do with a demonstration of national strength to all the world. Nevertheless, the doctors didn't consider Sepúlveda the right person for this first moment.

Also they had to resolve when to bring up the Bolivian miner, Carlos Mamani, who, as the only foreigner among the thirty-three, and in consideration of the diplomatic signals to that neighboring country, clearly should not be the last. The Chilean government could not feed any suspicions that Manami's rescue was in any way less important than that of the Chileans. So deciding on the first four to bring to the surface had turned into a very complicated task.

It took the authorities and the doctors several hours to settle a problem that, if it wasn't resolved soon, could have permanent and undesirable consequences for the image of the entire mission.

Finally, after a long conversation in which, according to Dr. Díaz, the impasse had been very hard to resolve, the doctors and the politicians reached a consensus.

Once again the greater good prevailed, and everyone conceded something.

The group decision was made. The first to emerge from the depths would be Florencio Ávalos, the fittest candidate according to medical condition.

Then would come Mario Sepúlveda, so that his enthusiasm could reflect the good cheer of those below. Even more importantly, his meeting with President Piñera would be a potent signal of optimism for the many millions of people who would be following live, from all parts of the world.

"ME LAST"

Juan Illanes, another of those deemed "strong," was the third on the definitive list, and after him would be Carlos Mamani. The president of Bolivia, Evo Morales, had recently arrived from La Paz.

The public wasn't aware of all this. Only the goal of the rescue mattered, and the lives of those at the camp revolved around that. The women combed their hair and retouched makeup and nails over and over. The men chain-smoked and debated the best strategy to follow so that things would turn out the best possible way, each offering his insights according to his experience and expectations.

Women, men, elders, and kids tried to fill these momentous last hours with whatever banal pastime occurred to them in the moment, and would no doubt be forgotten tomorrow.

While aboveground they were setting the order of extraction, several miners had expressed the desire to be the last to go up to the surface, giving up their place in line to workmates who were older or of more delicate health.

The minister commented, looking at the dust that covered his shoes, "They keep up an admirable spirit of solidarity, of comradeship. And there is no doubt that they have faced difficulties, but it is amazing how they have managed to keep such an enviable attitude, which we all can only admire."

Amid these spontaneous displays of altruism, it was decided that the last worker to reach the surface would be the surveyor Luis Urzúa—meaning he would spend the most time in that hot, dank stone prison. Like the captain of a ship who is the last to abandon the vessel before it sinks, the shift boss would wait until all his men were safe before finally caring for his own well-being. Some of his men wanted his stay inscribed in the *Guinness Book of World Records*. All his family, meanwhile, grew more anxious than ever, for until he was safe on the surface, nothing was certain.

Urzúa's mother, Nelly Iribarren, could not contain her tears: "I was terribly upset," she said, "like any mother would be, and I cried a lot. But now I'm happy. I'm very proud that this is about to happen. Soon he will have me for his mother again—Mother Earth, who has kept him all this time, will have to give him up."

Those close to Luis agreed that he was a born leader, and they were sure that he would never leave the mine until he had made certain that the rescue was complete. "He worries a lot about his people, and since the beginning he has guided the miners, giving them courage," said his cousin, José Astorga.

In Chilean mining, the tour foreman has implicit obligations and commitments that are quite strict, which helps explain the trust that everyone placed in Luis.

THE LAST MEAL

Once the inevitable bureaucratic meetings were over, the doctors concentrated on preparing the thirty-three for the most important trip of their lives.

The menu for what would be the last meal they ate before getting in the cage was now ready: a little salt, two scoops of French fries, and an imported energy drink that elite international athletes drank.

The intent was simple and vital: The men needed to stay as well-energized as possible during the trip, because they didn't know now long it would take each man to get to the surface. The specialists worried that someone might suffer a blackout in the narrow escape cubicle. So all the food was chosen to keep their blood pressure up, almost to hypertension, during that time and keep them metabolically balanced if something should go wrong.

RESCUE PREPARATIONS

The hours crawled by. Down in the refuge, the miners had come to the final part of the physical training they had been doing throughout the past weeks to prepare them for the ascent up a tube of 2,296 feet.

The medical team sought to guard against any eventual problems: effects on the lungs, bodily fatigue, or psychological stress. There were no precedents or international studies that addressed the situation that had occurred in the San José mine, or the physical and psychological conditions of workers who endured confinement underground for more than two months. There was simply no medical bibliography they could consult to compare similar situations—because there had been no similar situations.

"All this is new," noted Jean Romagnoli, the physician in charge of the group's physical conditioning. Both the miners and their doctors were on uncharted ground.

This doctor, a specialist in sports medicine, was a burly, stout man with the body of a rugby player. He was tasked with getting the miners in the best physical shape possible before their arrival on the surface. Romagnoli explained the process of preparation he had developed to get the miners ready for D-day.

"We know that the heat and humidity in the refuge work against the mission," he said. "We're also aware that a slip-up at the last minute could mess up everything that we've worked so hard to accomplish."

So *what should I have done differently?* Romagnoli asked himself, surprising himself with the question. But immediately, he waved it off and began to explain the physical preparations that he been transmitting for several weeks now to his trapped patients. He leaned forward and continued. "It was necessary to do tests to learn the cardiac response of each one, the lung resistance, even the psychological stress, and also to study other experiences internationally that were in any way similar. Then we designed an exercise plan based on a preventive model."

He shut his eyes and ran his hand over his face, wiping away the sweat beneath the safari hat that he always wore to protect his bald pate from the sun.

Romagnoli spoke at length, calmly and with austere pride. He said that the training underground had begun five weeks ago. Everything was supported by a video that recorded it all.

In the middle of a part of Camp Hope reserved only for the professionals on the rescue team, away from all other external distractions, Romagnoli paused to reveal details of the educational video put together for the miners. He was the one who demonstrated all the exercises on the video.

"My original idea was to get a model a good deal more attractive than I, to motivate them to do the exercises shown in the images," he explained with grin. "Well, what model wouldn't be?" he added, chuckling. "I couldn't get one, though. So it's just what I could do on my own," he said, pointing with both hands at his chest.

CHAPTER THIRTEEN

LIKE FIGHTER PILOTS

The training that Romagnoli imparted to the miners was known as "L1-Modified," the same exercise routine practiced by fighter pilots. It basically helped raise the blood from the lower extremities to the trunk, in order to avoid any sort of faintness caused from being in a rigid position.

To this were added cardiovascular exercises, which burned fat and made the miners weigh a bit less, since the cage that would lift them to the surface was only 21.25 inches wide.

Meanwhile, the miners were working with elastic bands to strengthen certain muscle groups, using distinct parts of the body, such as their arms and buttocks.

For everything to function well—or almost everything—Romagnoli stayed in permanent communication with the miners by telephone or videoconference. When necessary, they exchanged letters with vital information and comments for fine-tuning the instructions he sent from the surface.

"This interaction helps a lot in keeping up the group's state of mind, and in staying abreast of what's new in the refuge. This is fundamental because it's as if I were down there with them, in their environment, listening to their words, their needs, their ideas. There is a lot of relief on their part and a lot of support on ours," he said.

Monitors recorded the training sessions in the depths of the mine. "Then they send them up here and that way we can have the clearest view."

A CURIOUS REACTION

Even with the confidence Jean Romagnoli had cultivated among the group of thirty-three, some of the miners were resistant to doing the exercises suggested by the doctor. They didn't like the squats, push-ups, and stretches.

Some were just shy, and others felt embarrassed, as if the series of exercises recommended by the doctor put them in a position that was somehow undignified or humiliating, inappropriate for tough northern miners accustomed to hard, exhausting work among tons of rock and explosive blasts.

Although most of the *viejos* were willing to submit to the series of exercises imposed from above, some were not so willing. They hid in the twists and turns of the mine; they isolated themselves. Indeed, at times, the refuge seemed full of rebellious schoolboys rather than tough, experienced men.

But Romagnoli took an attitude of restraint before such pride and ego and tried not to attach too much importance to the behavior of the thirty-three.

"Some have revealed their inner selves for personal reasons, which is understandable in the context we are in. Especially," he added in an academic tone, "because the dynamics implemented at the beginning of the medical preparation were a little bit strange— pretty paternalistic, to be precise. They told the miners what they had to do, what they had to eat, how much water they could drink, and never asked for their opinion. This provoked some to rebel." Romagnoli spoke as if revealing a secret.

"When I arrived," he said, settling back in his chair to recount the experience, "the first thing I did was tell the miners, 'It's like this: I have the impression that the system of medical attention isn't functioning all that well, and just so things can go well for us, it should be up to *you* guys to follow the instructions from topside. And at the same time, you have to participate in the decisions. . . . You, too, have your opinions, and we should be listening to them.' And just like that, I gained the trust of them all."

THE SPOKESMAN IN THE REFUGE

For the success of his work, Romagnoli relied all along on the special collaboration of a "direct contact," as he liked to say.

He was referring to Mario Sepúlveda, who had turned into a sort of spokesman and intermediary for his comrades.

"Mario is a great help with the motivating force that he manages to convey to his colleagues. He's a chin-up kind of guy—he's the glue that holds the group together," Romagnoli said with pride; later the two would become great friends.

ALMOST READY TO GO

Since the moment the miners could see the drill head of the T–130 above them, they had started getting ready for their impending return to the surface. Gaping like wide-eyed kids, they let it sink in that, after the seemingly endless night of waiting and more waiting, what they had been hoping and praying for was finally about to come true. After weeks of strenuous physical training, now it was time for the medical tests to predict how they would react physiologically during the ascent in the capsule.

Meanwhile, up on the surface, a crew of workers began dismantling the drill that had made history, as pieces of the machine that would hoist the capsule to the surface began arriving for on-site assembly. And a group of medical experts continued testing and perfecting the thirty-three biometric belts that would be used to monitor every second of each miner's claustrophobic trip up.

"It's the kind the NASA astronauts use," explained Ben Morris, the American engineer representing Zephyr Technologies, the company that manufactured the biometric measuring devices. The thick, heavy belts, costing two thousand dollars each, could measure skin temperature, heart and respiratory rates, blood pressure, and oxygen consumption.

According to plans, all the data would reach the medical team on the surface in real time, using a Bluetooth connection. In those final moments, technology merged with ideas, ideas with experience, and experience with faith, to pull off a rescue viewed by many as nothing short of miraculous.

DRESSED FOR THE TRIP

The jumpsuits the thirty-three wore for their escape were a story all their own. They were made of Hipora, a waterproof, windproof, breathable fabric that would keep their skin dry. Underneath, they wore special antifungal t-shirts.

The entire rescue operation would adhere to a strict protocol. An action plan had even been devised in the event that one or more of the miners died during the process—an unpleasant but very real possibility. The doctors on the scene hoped with every fiber of their being that they wouldn't have to follow that grim directive, though they planned for it nonetheless. Nothing in the operation could be left to chance.

The moment each miner arrived on the surface and left the capsule, Dr. Andrés Llarena, the head doctor of the Chilean navy, would receive him. Then, after a quick examination, the doctor would pass him along to Dr. Lilian Devia in the camp's medical triage area. Just a few hours before the grand task began, Dr. Devia joked, "This is going to be like birthing sextuplets—except there will be thirty-three of them!"

Dr. Devia and her entire team would remain in the emergency medical area until the last miner had been pulled to safety. She explained that the miners would arrive on stretchers. The triage tent was equipped with resuscitators, heart monitors, equipment to administer anesthesia, and drugs to counteract cardiac arrest. No matter what their condition when they arrived, the men would stay there for at least two hours of observation, to make absolutely sure the medical team detected any physical harm they may have sustained en route to the surface. Two family members of each miner would be allowed to stay with him, as long as he was not exhibiting any serious health problems, until the transport by helicopter to Copiapó. Each miner, without exception, would stay in the city hospital there for at least forty-eight hours. What happened after that would depend on the miners' family and friends, who had all been through an intensely agonizing journey of their own aboveground.

In one of Copiapó's neighborhoods, known as TilTil Bajo, all the residents anxiously awaited the return of Carlos Bugueño and Pedro Cortez, two of their neighbors who had gone to work at the mine that morning of August 5 and hadn't been seen since. *"Welcome Home!"* read the banners and posters that had been plastered on every front door in a show of solidarity.

"We've cried and prayed for them, and now we're ready to bring them home," said a neighbor who had known both miners since they were children. "We've been sending letters to their families, and they write back telling us how much our support means to them."

The whole neighborhood kept a vigil at the local chapel, with one eye on the televisions that broadcast nonstop coverage from the San José mine. When the two men finally came home, the neighbors were going to block all the streets off and have a huge party. A group from the church had thoughtfully planned out what would be the most heartwarming reception their congregation had ever witnessed.

"We're so proud of them," their old neighbor commented. "Now everything at the end just has to go right." In just twenty-four hours, the journey no one would ever forget was about to begin.

THE RESCUE WORKERS GEAR UP

All alone, not far from the little city that Camp Hope had become, the team of professional rescue workers who would bear the enormous responsibility of actually getting the miners to the surface went about their preparations with an air of almost disconcerting calm. They knew that if any miner did not make it, they would be the target of the most vicious recriminations. But they also carried with them the calm certainty that they were fully capable of pulling off their daunting mission.

Among them was Manuel González, forty-six years old and one of the most seasoned of the group, having spent twelve years as a mine rescue specialist. He had extensive experience in Codelco's El Teniente copper mine, the largest underground mine in the world, located south of Santiago. His peers saw him as a consummate professional in excellent physical condition, skilled at directing groups of workers confidently even under the most dangerous conditions.

In his younger days, González had been a professional soccer player, but he traded in the lush green playing fields that earned him more goals than money for the dark, gloomy underground landscape of El Teniente. The sacrifice came with a healthy paycheck. Now he was perfectly at home working in subterranean tunnels, digging out of collapses, surrounded by danger and the head-splitting din of machinery. He was proud of his new career.

Now up north at the San Jose site, González had pored over the videos of the mine's geological condition, preparing for the rescue proper, which was fast approaching. He felt thoroughly confident in his ability to do the job, and he was ready to get on with it—or at least appeared to be. He and his fifteen crew members were all set to begin the final phase of rescue efforts to bring the thirty-three to the surface. Ten of the rescuers worked at the National Copper Corporation (Codelco), three were medical personnel from the navy's submarine fleet, and the rest were local emergency workers from Atacama, who would pull alternate twelve-hour shifts.

Six men would go down into the mine, while the rest would play supporting roles aboveground. Although they hadn't been chosen yet, Manuel González was convinced he would be one of the six making the journey into the depths. In spite of his confidence, he didn't say anything. Fellow rescuer Patricio Roblero, brought in from the navy, was far less circumspect about his chances of being chosen. He loved taking on the hardest missions, and he wore his patriotic spirit on his sleeve. "The first thing I'm going to say when I get down there is '¡*Viva Chile!*'" Roblero declared. Keenly aware of the rescue's historical significance, he had scripted out in his head exactly how it should unfold. Meanwhile, the officials overseeing the entire operation were busy readying the hoisting equipment, so that they could welcome the first miner back to the earth's surface before the day's end.

Once again Chile's minister of mining, Laurence Golborne, stood in front of a sea of cameras. Running a hand through his perfect hair and adjusting his official bright red jacket, he lowered his head and waited for the photographers' signal to begin. Then Golborne paused a beat before delivering the news that would then fly around the world at electronic speed.

"The long-awaited rescue operation will begin at midnight on October twelfth," he announced in a clear, self-assured tone.

Suddenly the tide of reporters surged forward, pushing and shoving. Power cables snarled and tangled as cameramen, too, got caught up in the frenzy. Some cameras were even smashed and trampled in the mad dash, but there was no time for apologies. Back in television studios around the world, the anchors were ready to turn the coverage over to their reporters on the ground, who were hastily straightening their hair, tucking in their shirttails, and doing their best to camouflage the fatigue on their faces and collect their thoughts.

The entire camp seemed to hold its breath. The very air was charged with all the emotion that had been building for over two months now and was about to burst forth—but not quite yet. There was an almost palpable feeling of faith as the desperately longed-for moment finally arrived.

One of the happiest in that expectant crowd was Francisco, the thirteen-year-old son of the miner Mario Sepúlveda. After the engineers' last meeting with the family members, Francisco stood outside the tent with André Sougarret, admiring his shiny red hard-hat. "Do you like it?" he asked the boy, who responded with a shy, delighted smile. "When the rescue's finished, I'm going to give it to you," he promised.

The number of journalists at the camp, only 300 a few weeks ago, had ballooned to over 2,500. The camp was literally overflow-ing with people, emotion, and suspense. The tents of the media out-lets had become a city, spreading out on the desert sand and stone, dominating the landscape. Some of the miners' family members com-plained about the media's constant badgering, lamenting that the res-cue effort had turned into some kind of made-for-TV reality show.

Golborne responded to their collective concern: "A lot of peo-ple are saying that this feels like a reality show," he said, "but the reality is, this has affected the whole world. Hotels are filled to capacity, and we have over two thousand journalists here from all over, so as the government and state, we have to manage it. We didn't ask for this, but this is the situation we are faced with. We need to deal with it and project a positive image of our country, and make sure the visitors have a place to stay, that there are enough bathrooms, and that they can set up their news vans for satellite transmission. We have to plan for that, too, because it's part of the respect they deserve from us, and the image we're going to project as a country in a situation that, whether we like it or not, is going to be closely watched by the rest of the world." A firm message, but one delivered with his characteristic good humor and tact.

CRASS COMMERCIALISM

While the minister of mining answered reporters' last-minute questions, an uncommon scene was unfolding on the platform that had been erected to receive the rescued miners. In a tactless attempt to exploit the marketing power of all those camera lenses from around the world, executives from a well-known sportswear brand had managed to send some items of clothing for the miners to wear as they were being pulled out of the mine. Polo shirts, pants, and white running shoes all bore an instantly recognizable logo that at the critical moment would be seen by millions of television viewers all around the world.

The marketing executives gave their best pitch to some government officials, who, in the end, vetoed their efforts to commercially brand the rescue. One of the miners was happy enough to take the gift, though he wasn't quite sure what he could do with it. He commented, "It's great to get a free pair of shoes, but I'm a size forty-two, and these are all thirty-eights."

THE RESCUE IS PRESIDENT-PROOF

As the countdown ticked away, daily life in the camp came to a standstill, as if everyone were collectively preserving their strength for the monumental event that was fast approaching. At noon, three family members of each of the miners assembled at the security barrier, around the hoist that would bring the capsule to the surface, so that they could be there when their loved ones stepped out of the scariest, most claustrophobic elevator ride of their lives. At the same time, reporters and camera operators put on the official bracelets that would allow them access to a designated area 230 feet away from the capsule, where they would be able to take pictures and record the drama the whole world was anxious to see unfold.

In the afternoon, it was time to run a final series of tests on the equipment. Miguel Fortt, one of the engineers on the scene, explained that for purely aesthetic reasons, they had decided to use the second of the three capsules, the Phoenix 2, because it looked the cleanest. It was painted the red, white, and blue of the Chilean flag. Unlike the first capsule, which was simply a heavy tube, the second capsule had a door that closed with a latch. So when President Piñera symbolically shut the capsule door for its first trip down into the mine, he didn't do it properly, and the latch jammed. The problem was quickly fixed. Fortt remarked that of course the president wasn't an engineer and shouldn't be faulted for not operating it perfectly. "It wasn't an error on the president's part," he explained to an insistent reporter. "We're not trying to put the blame on anyone or look for conflict in what has been such a flawless and successful operation. It's an accident that could have happened to anyone at such an emotional time. But we have to let Asmar know that next time they'll need to build a capsule that is president-proof," he joked.

DOWN TO THE WIRE

In just a few hours, the rescue would begin in earnest. Government officials would wait until the very last minute before revealing the names of the six rescue workers who would go down into the mine.

As the minutes ticked by and a blanket of stars spread across the desert sky, Manuel González and his fellow rescue workers headed toward a trailer that had been set up just a few yards away from the rescue platform. It was far enough from the mass of reporters to afford them some privacy. The trailer had been equipped with beds for each of the men, in case the mission should take longer than anticipated and they needed to get some rest. Time seemed to stand still as the stars sparkled across the clearest skies on earth.

In Camp Hope, the expectant tension was almost too much to bear. Finally, the authorities announced the six rescuers who would go down in the capsule: Sergeant Roberto Ríos and Corporal Patricio Roblero, both of the Navy special forces; Jorge Bustamante Ramírez, from the state-owned mining conglomerate Codelco; Corporal Patricio Sepúlveda of Chile's elite Special Police Operations Group; Atacama firefighter Pedro Rivero; and Manuel González, the rescue specialist from Codelco. After the names were read, González's face betrayed no emotion. Perhaps his steely reserve masked a concern over the extra responsibility he would shoulder as the first rescuer to descend into the mine, and the last man to emerge from its depths.

CHAPTER FOURTEEN

THE BEGINNING
OF THE END

Now, as the rescuers began their mission, there was nothing left for everyone else to do but watch and wait. Since
everything was in place, they decided to begin a few
hours ahead of schedule. At long last, the task of rescuing the thirty-
three men who had been trapped deep underground for sixty-eight
days could begin. The final chapter was about to be written about a
group who, since that fateful day of August 5, had nurtured a stead-
fast faith that they would be brought back alive from the dark tomb
where, together, they had wept, prayed, joked, argued, and planned
for this day. They had somehow managed to pull together and survive the kind of disaster that rarely offered a second chance at life.

Up on the surface, dozens of people had gathered on the platform that would serve as the reception stage for the Atacama
thirty-three: doctors, psychologists, nurses, engineers from the
federal and regional governments, news photographers, family
members of the trapped miners, and the team of rescuers who

were ready to go. Each had a crucial role to play, and now they awaited their cue to act.

All eyes were on Manuel González as he climbed into the Phoenix 2 capsule at 10:10 P.M. "Let's go, Manolo!" the other members of the rescue team shouted while President Piñera looked on. Then, as if seeking reassurance, Piñera smiled at his wife, First Lady Cecilia Morel, who also happened to be a professional family counselor with over thirty years' experience. She had spent a great deal of time with the trapped miners' family members, especially the women. She stood with her head lowered, as if in prayer.

Everyone started clapping at once in a spontaneous show of support for the rescue team. Stepping into the capsule that was just barely big enough to contain him, González wore an orange jumpsuit and a white hardhat with a built-in headlamp. A camera installed inside the Phoenix 2 would film the entire operation. The door was latched from the outside, and Manuel González began the slow, steady descent into the abyss. "Just relax, Manuel, like you're lying on the beach," one of his colleagues shouted after him as the capsule disappeared from view. As the Phoenix 2 was lowered down the dark shaft no wider than a bicycle tire, Manuel's main concern was that the retractable wheels on the outside of the capsule not get hung up along the way. A few minutes into the descent, workers guiding the capsule downward paused it to fix a guide to the cable attached to it, to help keep it centered in the shaft. Then the Phoenix 2 continued its downward journey, every detail under close scrutiny. Meanwhile, on the surface, President Piñera and the other officials assembled around the hoist erupted into a rousing refrain of the Chilean national anthem. Then finally, at 10:37, the Phoenix 2 touched down in the mine's interior.

Everyone watching the action on any of the four giant screens that had been set up in Camp Hope saw Manuel González step out of the capsule, his back to the awaiting miners. The crowd above erupted in a roar, some sobbing tears of joy, thanking God at the top of their lungs. At first, González was on the screen alone, though he was quickly joined by three miners, and they all embraced in front of a Chilean flag that had been hung up beside the Phoenix 2's landing point. For many, watching González step from the capsule out into the surreal landscape of the mine brought to mind the lunar landing of July 21, 1969, when Neil Armstrong became the first human to set foot on the surface of the moon. And in many ways, this rescue mission, mounted against such grave odds, was also a "giant leap for mankind."

The deafening applause up on the surface continued as González stood with his hands on his hips in the steamy darkness and described for the thirty-three what going up in the capsule would be like. Many of the miners were dressed only in shorts, their torsos bare. Unable to contain their joy, they hugged and kissed González over and over again, as if he were an angel sent down from heaven above.

González explained that he would be the very last man to leave the mine, after everyone else had been brought safely up to the surface. Motionless, the group solemnly listened to their savior sent down from above, until González suddenly lightened the moment: "You have no idea what a wussy little shit they sent down here to get you. Now, there are tons of people up there waiting for you, so let's get outta here." The group burst out laughing, and the tension evaporated. Eager to get moving, they sprang into action and quickly prepared for the first miner to leave.

Days ago, it had been decided that thirty-one-year-old Florencio Ávalos would be the first up, since he was in the best physical
condition. He put on the special belt to monitor his cardiac and
pulmonary activity during the ascent in Phoenix 2. He followed
the instructions precisely, not wanting to risk the slightest error. He
knew he was about to get his second chance at life.

CHAPTER FIFTEEN

FREEDOM BOUND

The minutes ticked by, and on the ground everyone looked positively ecstatic over how the rescue operation was unfolding. Florencio Ávalos was on his way up. The rescue was the only thing being watched on television throughout Chile, and in many other countries as well. Gathered at the hoist where the capsule would emerge, President Piñera, doctors, engineers, and various government officials waited nervously, collectively holding their breath, though it wouldn't be for long. Florencio Ávalos's seven-year-old son, Bairon, overcome with emotion as he waited for his father to emerge, sobbed, and his tears touched the hearts of everyone around him.

It was 12:11 A.M. on October 12, 2010, the date that, 518 years earlier, Christopher Columbus first reached the Americas. At that moment, the Phoenix 2 reached the surface, and Ávalos, who became the number two man in the hierarchy of the thirty-three, stepped out onto the ground sixty-eight days after the earth swallowed him up. Right away he kissed his son and pulled him close, hugging him tight for a long time, as if neither one ever wanted to let go. Then he hugged his wife. "Thank you," he said simply—the only words he spoke, and they were enough.

On the t-shirt he wore over his jumpsuit, the words "Thank You God" were written over Chilean red and blue, carefully signed by each of the thirty-three. In the Chilean government's official television coverage of the rescue, a small inset screen showed the miners still underground, now more anxious than ever to get out. Once Florencio had successfully reached the surface, everything moved quickly. The second rescuer, Roberto Ríos, immediately climbed into the Phoenix 2 and was on his way down into the mine.

The rest of Florencio's family watched the rescue from their tent in Camp Hope, on the screen of a small television that had been set up for them. As they watched their loved one step from the capsule, their hugs and cries of joy were quickly overshadowed by the frenzy of reporters and cameramen jostling each other for a position right outside the tent. They frantically shouted questions at the family, many in foreign languages. The scrum of journalists was so out of control that they inadvertently knocked loose the tent pegs, and the tent slowly started to collapse. Everyone yelled, no one listening, no one able to be heard. The near riot caused by the media frenzy was the one black cloud hanging over the rescue—though many people had predicted it, no one could prevent it.

Down in the mine, the rescue proceeded apace. Manuel González talked to many of the men, who were now restless, having trouble holding it in after months of waiting now that rescue was imminent. He tried to exert a calming influence and not let the group's collective energy get too agitated at this critical hour. And everything unfolded exactly as the team of engineers and experts had painstakingly planned. Now a group of thirty-two, the men listened attentively to their instructions and then dispersed. The next few to be hauled to the surface waited by the shaft, while the others went off and stretched their legs, moving from one side of the space to the other, trying to relax in preparation for their turn in the capsule. The two rescue workers called the names of the miners in the order they would go up. As a name was called, the chosen man went over to the improvised chapel the men had set up with lit candles. And there, alone, he offered a final prayer to his God, drawing on the faith that had played such an important part in sustaining the whole group. Then, before climbing into the Phoenix 2, each miner took a moment to gather whatever few small things he might want to bring with him: a rock, a letter, a spoon. Then he slowly stepped toward the capsule, drawing in his last few breaths of the mine's thick, sweltering air.

The second man to be rescued, Mario Sepúlveda, made a memorable entrance back onto the surface. As soon as the Phoenix 2 lurched to a stop, he yelled at the technicians standing by, "Get me out of here!" The second the door sprang open, he ran into the arms of his waiting wife, Elvira Valdivia. They clung tightly to each other for the longest time. He tried to say something to her but was too overcome with emotion to speak. The helmet fell from his hands, landing in the dirt. Finally he managed to say, "We have the rest of our lives together to talk . . ."

Mario had a small sack slung over his shoulder. Taking it, he reached inside, looking like a bald, skinny Santa Claus of the desert, and started handing out rocks gathered from the mine. When he hugged Minister of Mining Laurence Golborne, he blurted out, *"¡Puta! ¡Jefazo!"* (Holy shit! The head honcho!) When presented to the First Lady, he kissed her hand gallantly and said, "Madame, it is an honor."

Euphoric, the supposedly weak and exhausted miner then ran in front of the gallery of observers and raised his arms. "Okay kids!" Mario shouted, "Chi-chi-chi . . ." Leading them in the patriotic chant, he pumped his fist as his stoic compatriots joined in, many now blubbering like little children. Finally, the medical personnel managed to convince Mario to get on a gurney and be taken to the triage area, where he continued to entertain. "We can stay here tonight," he said, pointing back toward the mine. "I made my bed very nicely this morning, just in case."

The rescue continued at a fast clip. With half the men already on the surface, Omar Reygadas would be the seventeenth to get out. He was a nervous mass of pent-up, conflicting emotions. Later, he would admit that before stepping inside the capsule, he looked back almost wistfully one last time at his bed and the little make-shift desk where he had written so many letters to his family. Those things would stay behind forever. "I really wanted to leave, but I also felt kind of nostalgic leaving the things that had been with me down there for all that time," he said.

After Reygadas, the other miners behind him quickly made their way up, carefully adhering to the established protocol. It was a whirlwind of cheers and applause, emotional reunions, and heartfelt speeches as night gave way to dawn, the fog keeping a respectful distance on this momentous morning. The group's leader, Luis Urzúa, was the last miner to reach the surface. Twenty-three hours had passed since the historic rescue operation got underway. Urzúa hugged the rescue workers and President Piñera, who had been standing by the hoist, almost without interruption, the entire time, except for a few minutes when he went to the bathroom and changed his shirt.

It was an electrifying moment. Many of the rescue workers, medical personnel, engineers, and other support people who had been working tirelessly to see it come to fruition now hugged and kissed anyone within reach, so thrilled and proud were they of what they had managed to accomplish together.

Calm and collected, not showing any visible signs of strain, Urzúa talked with the president of his country. "We had the strength and the faith to fight for our lives," he said simply. Both men looked each other straight in the eyes, speaking clearly, each aware of how pivotal this encounter was after the ordeal they both had been through.

The head of state asked what the hardest time was down in the mine, and the leader of the thirty-three replied without hesitation, "There were several. For example, when the collapse happened and we saw the rock blocking the way out, a lot of us thought we would be rescued within a day or two, because the company had always said it wasn't dangerous. But we knew how to take care of ourselves. The first few days, we did some things that weren't very smart, but we stayed calm." He paused. "I hope this never happens again," Urzúa told the president as he symbolically punched out of his shift, which had begun on August 5, sixty-eight days earlier— the longest shift of his life.

WHAT TIME CANNOT ERASE

Since the day of the rescue, the miners have carried on with their lives, following their disparate, unpredictable paths. After staring death in the face, they have been born again, to discover new talents—and, for some, also character flaws—they hadn't known they had. Now they are seizing this new chance at happiness, fully aware of how precious the simplest pleasures in life are.

Most of the men have taken quiet refuge in the warmth and comfort of their families and homes. But a small number of them who captured the media's insatiable attention have learned to live a life under the glare of the spotlight, vastly different from their previous existence spent toiling in anonymity.

For example, Yonni Barrios, famous for his soap-opera story of infidelity, was offered $100,000 to appear in an advertising campaign for AshleyMadison.com, a controversial dating website for married men looking to have affairs. He also received a substantial offer from a company marketing a drug to boost male virility in Chile and Central America. He turned them both down.

Edison Peña, now known the world over for his fanatical devotion to Elvis Presley, traveled to New York and appeared on the *Late Show with David Letterman* on November 4, 2010, getting a warm reception and lots of laughs from the studio audience. Although he primarily talked about his experience of being trapped and rescued from the mine and he had to rely on an interpreter sitting next to him, his quick wit and charisma made for a really fun and inspiring interview. And the highlight of the segment came at the end, when Peña leaped out of his seat and burst into a rousing rendition of "Suspicious Minds," complete with Elvis dance moves, as the show's bandleader, Paul Shaffer, played along. The audience roared their approval for the night's biggest star.

And that was just the beginning. That same week, Peña ran the New York Marathon. At the race's starting line on the Verrazano

Bridge in Staten Island, he greeted New York City mayor Michael Bloomberg, and the elite runners gathered around him. Wearing the number 7127, he joined the throng of 45,000 runners from over 100 countries around the world. Running alongside him the whole way were his two escorts, Juan Jesus López and Rene Cuahuizo, who got their own moment in the spotlight with articles in the *New York Times* and other papers telling the story of how they came to accompany the world-famous miner. Both hardworking immigrants from Mexico and New York City residents, López and Cuahuizo were experienced runners who had done the marathon before. But in 2010, the entry fee was too steep, so they had not entered. That all changed when Peña accepted the New York Road Runners Club's invitation to run the race. Then Spanish-speaking escorts had to be found to keep other runners from crowding Peña and to translate along the route if needed. So when Peña crossed the finish line five hours, forty minutes, and 51 seconds after he began, it was López and Cuahizo who held the Chilean flag aloft in his wake, cheering him on as his waiting wife, Angelica, proudly received her husband. Peña's newfound fame took him down to Wall Street the next morning, where he stood alongside the winners of the marathon and rang the closing bell at the stock exchange, bringing an end to the day's trading. To top it all off, on January 7, 2011, Peña found himself in the holy land: Graceland. He was a special guest for the annual Elvis birthday party, marking what would have been the King's seventy-sixth birthday.

All of the thirty-three got something in the way of perks following their rescue: They all received an invitation to go to Disney World for an all-expenses-paid six-night vacation with their families. Each family received a five-hundred-dollar gift card to spend however they chose at the park, and the miners were invited to serve as grand marshals in the parade that wends its way down Disney World's Main Street every afternoon.

Going in a more practical direction, the rescued miners Omar Reygadas and Mario Sepúlveda tapped into their inner strength and faith in themselves to launch careers as motivational speakers, captivating large audiences with their seminar entitled "Trapped: With a Way Out. You Can Do It." They say they are well paid for their presentations, and they get more and more offers every day. Addressing a group of college students in southern Chile, both survivors spoke about how we can use the power of the human will to achieve great things and bring about social change and, in doing so, take control of our own lives. Accompanying the two miners is another partner in the fledgling business, Dr. Jean Romagnoli, the doctor overseeing the miners' fitness training down in the mine. Romagnoli speaks about human response in extreme situations and describes what it was like working with the team of NASA technicians during the rescue.

Mario Sepúlveda has capitalized the most on the miners' worldwide fame and media attention, thanks to his naturally extroverted personality and contagious enthusiasm. In less than three months after the rescue, he visited the United States, Germany, the United Kingdom, and Argentina, accepting invitations from international organizations and media outlets. He then was off to Japan, New Zealand, Russia, Israel, Puerto Rico, the Dominican Republic, Greece, and Africa. The earnings from his motivational seminars are managed by his wife, Elvira Valdivia, an accountant by profession. Back home in Pudahuel, one of the poorest sections of Santiago, Mario was proudly recognized as an "Illustrious Son" of the area and presented with a key to the city.

On a more romantic note, five of the miners officially proposed to their girlfriends at a party in Caldera, a coastal city near Copiapó, in the first public gathering any of the miners participated in after their dramatic rescue. Esteban Rojas was among them. He promised to finally marry his wife, Jessica Yáñez, in a proper church wedding, twenty-five years after they were wed in a simple civil ceremony. He first delivered her proposal to her in a note sent up from the depths of the mine in September. At the party, hosted by Chilean impresario Leonardo Farkas, Esteban publicly announced his plan to marry Jessica before the eyes of God, in a church filled with family and friends. "I accept. I still have the note!" Jessica said, overcome with emotion.

But for others, their newfound freedom is a precious commodity that must be vigilantly protected, in an effort to carve out some much-needed privacy, peace, and quiet outside the glare of the spotlight. That's how it is for the group's shift leader, topographer Luis Urzúa, the man who maintained discipline and fostered a spirit of camaraderie during their long ordeal. He and his family have politely refused all requests to talk to the media. Urzúa's top priority now is to closely follow the investigation being conducted by the Atacama district attorney to determine who bears legal responsibility for the accident at the San José mine. Urzúa often meets with Héctor Mella, the attorney in charge of the investigation.

While most of the thirty-three try hard to go back to the lives they had before the mine's collapse, a sense of normality eludes many of them as they grapple with the demands of the press, public appearances, and a sea of mixed emotions that can be psychologically troubling. A sudden stroke of good luck on the scale they experienced can sometimes affect one's emotional equilibrium, and that has been the case with several of the survivors.

Many of the men have participated, with their families, in intensive therapy sessions designed to help them manage and overcome post-traumatic stress. According to the Chilean government's medical team in charge of the miners' ongoing treatment, six of the men began drinking heavily in response to all the emotional turmoil.

While the men were down in the mine, they made a pact together that they would not speak to anyone outside their closest family about their ordeal. But the need to talk about traumatic experiences can be overwhelming, and sharing difficult moments with another can be cathartic. The heart can't keep things locked inside forever. And that is why two of the men, José Henríquez and Víctor Zamora, opened up to me in exclusive interviews for this book. Following are the private conversations we had shortly after the rescue.

CHAPTER SIXTEEN

GOD IN FLESH
AND BLOOD

A few days after the rescue, José Henríquez felt he needed to travel to his hometown in Talca, in the south of Chile. There friends, neighbors, and members of the congregation of the New Horizon Evangelical Church, which he had belonged to for decades, anxiously awaited him. Once back home, wanting to do nothing at all besides rest, he found his days filled up with one public appearance after another—something that had not figured into his plans at all. Everyone wanted to see him, talk to him, touch him, embrace him. The much-longed-for days of peace and quiet at home with his family would have to wait.

The honors and accolades came one after another. In San Clemente, the village where he was born, Henríquez was recognized as an "Illustrious Son." Then on October 27, he was off to Valparaiso for the National Day of Evangelical Churches, to be honored along with his fellow evangelical Christian miners Osman Araya and Omar Reygadas. He had to accept all the invitations from local authorities, sports clubs, social groups, and old friends. It seemed impossible to turn down any of them.

When I met with him after those frenzied days in October, he was trying his best to relax at his home in Talca, a small city 155 miles south of Santiago, where life proceeds at a slow pace, in a natural landscape utterly different from the northern desert. He had spent several tranquil days at home with his wife and twin daughters, Karen and Hettiz. Henríquez shunned the media and their frantic attempts to wrest from the thirty-three their deeply personal stories of what happened down in the mine. Still, José Samuel Henríquez, age fifty-six, who had played the part of the group's spiritual leader, acknowledged solemnly that in his heart, he wanted the message of Christ that was manifest in the mine to be made known.

José let me know that he could talk for only ten minutes, because he had an unavoidable family obligation. His wife, Hettiz Berrios, seemed nervous. She looked at her watch several times as her husband was about to open up about those endless dark days and nights spent down in the mine. Even though he allotted a very small time frame for our talk, José Samuel was clearly very willing to talk about the spiritual guidance he offered his fellow miners. He was certain that their all having withstood two months trapped underground was nothing less than "a miracle of God."

"I COULD HAVE DIED AT ANY MOMENT"

Henriquez confesses that just a few days before the mine collapsed, he received two signs that foreshadowed misfortune ahead. The first was a divine revelation his grandmother Sara shared with him. "My grandmother got a message from God," he says, "and twice she told my mother that I was going to experience a very difficult time, and it would be very hard for me to even get out of it alive . . . honestly, I didn't pay any attention to it."

The second omen came through his daughter Hettiz. On the day he was to travel north to go back to work in the mine, she arrived late to see her father off. It was very unusual, since she was always so punctual.

"She got there at the very last minute. I was just about to leave. I couldn't wait anymore, and then I saw her rush out of the car. She ran to me and gave me a really strong hug. It took me by surprise. She seemed upset. I looked at her closely and asked, 'Sweetie, why did you hug me like that?' 'It's nothing,' she said, but I didn't believe her. *Something's going to happen,* I thought to myself."

Aside from his family members' premonitions of tragedy, Henríquez himself experienced a harrowing experience at the San José mine, six months before the collapse. There had been a gas leak, and Henríquez had had to urgently evacuate two of his fellow miners. While trying to carry a third man to safety, Henríquez passed out. The close call left him with a nagging sense of dread. "I told my wife that I could die at any moment in the mine; because of the conditions there, it wasn't safe. When I said good-bye to my family, I left the house feeling that that something was going to happen to me."

Henríquez is quiet for a moment.

"What happened that day of the accident?" I ask.

"I was about to go off shift after performing some routine maintenance on one of the machines," Henríquez recounts. "As I was leaving, it happened. It was a huge noise, like an atomic bomb going off or something. All I could see was a big cloud of dust, so

I had to wait for that to settle before I could look around and see what had really happened. The way out was blocked, and there was nothing to be done about it. We were in a really extreme situation."

"How did you react?"

"Well, I just acted normally because panicking wasn't going to do me any good. When you walk with God, things have to work out; you have to keep looking for a solution to the very end."

"And what did the rest of the men do?"

"They calmed down and we started to get ourselves organized."

José is quite circumspect about how reason somehow won out over terror for the group, but then he offers me a very simple explanation.

"We asked ourselves what we had: air, and water from the storage tanks. So we had to get organized and plan how we were going to survive, because we had long lives ahead of us."

"Did you think you might get out of there right away?"

Slowly shaking his head, José answers with his characteristic patience: "No, there was no way for us to get out, either through an air shaft or somehow digging through a wall. We didn't have the tools, much less the strength to dig and climb out. There was nowhere to go. It was like finding yourself in a den of snakes . . . because without any lights, or anything, it wasn't possible. We couldn't say, 'All right, now we're going to do this, that, and the other.' But anyway we made lots of noise; we made a fire so that people outside might see the smoke that would escape through the air shafts."

"At any point did you think that maybe you were never going to get out of there?"

"Well, that was our first thought, because there was nothing we could do, but God . . . was always there with us."

"So you had to completely depend on the people aboveground?"

"Of course we had to depend on the outside, on whatever was starting to happen . . . since down there, like I said, there were just no options for escape at all. So accepting that, we started to get ourselves organized. We said, 'This is the reality and we just have to deal with it, and we have to give it our best,' meaning we had to keep ourselves alive and in our right minds."

"And you said that then just as calmly as you're saying it to me now?"

"Yes, that's right."

"But some of the others were panicking?"

"Yes, of course there was a lot of anxiety, not knowing what was going to happen, because effectively confronting the disaster right away was something only some of us could manage . . . and looking back now, I'm sure not everyone could. So, as the days went by, as a group we all faced the situation, and we began to deal with it."

His voice trails off, not finishing the thought, so as not to give away too many details, since before the rescue they all vowed not to talk publicly about their experience in the weeks before they made contact with the surface. But he does describe the hardest moment for him, when they went to open the box that hopefully contained emergency rations, kept in the safe room, which they were able to reach.

"At the beginning I told them: 'My friends, I promise you there's a church throughout the whole country praying for us. We're going to ask that whatever food that is within this box be blessed.' We said, 'O Lord, bless this food,' and I told the men, 'This will be multiplied; have faith; we are going to eat well; you'll see how happy we'll be when the blessings start to arrive.' That was one of the things I said to them, always focusing on the positive, always trying to cheer them up."

José recalls that at the beginning the group's spirits were very low, maybe in an even weaker state than their famished bodies. In stark contrast to the group, Henríquez faced the situation calmly and evenly, having already experienced the very worst kinds of tragedies imaginable over the course of his thirty-year career. For example, there was the time in 1986 when a dozen of his fellow workers were killed after a landslide buried the worksite in central Chile where he was working along with his father and his brother Gaston.

"What did you and the rest of the miners do right after the collapse happened at the San José mine?"

"Well, first we went to the safe room to take stock of what we had and what we didn't have on hand, so we would know exactly what we were dealing with," he recalls. "Then we started to make some decisions, to plan what we were going to do and see how we could keep ourselves occupied. We assigned tasks, figured out where we would sleep and what would be our living area, and we made a careful revision of everything we had, so we could weigh our options."

"What other aspects of the mine presented challenges?"

"The temperature was suffocating, since it was always so hot, around 95 degrees all the time, maybe even more. That was definitely the worst part. We sweated even when we were sitting still. We were really dehydrated because of that."

"How did you decide who was best suited to which tasks?"

"It was voluntary, so the group functioned democratically, by majority rule, or fifty-one percent. We threw out ideas, and if somebody didn't agree, fine; we were thirty-three people, and if we wanted to make a decision, we took a vote. No one made individual decisions, so we could all see the best option."

At this point in the conversation, his wife, Hettiz Berrios, interrupts. She points to her watch, trying to get José to wrap things up. She finally has her husband back, and, quite understandably, she wants him for herself. But José isn't in a hurry anymore. The ten-minute time limit we initially agreed on for the interview no longer seems to matter.

FAR BELOW THE GROUND . . .
CLOSER TO HEAVEN

"Did the stress the group was under affect your relationships with each other?"

"Sure. It started to get harder to get along. People have different personalities, you know. Some are stronger, others not so much."

"How did the group ask you to give them spiritual guidance?"

"They just asked me straight out, because I don't really make a lot of noise. I've always kept a pretty low profile. I keep to myself usually. So they said, 'José's an evangelical, so let's ask him if he would lead us in prayer.' I told them 'yes, I believe in a living God, and we're going to worship that God.' I said, 'if you all want to do it the way I want to, like the evangelists do it, then fine. If not, you should find somebody else.' I just put it to them very simply, but not being arrogant about it."

"And what did the rest of the group say to that?"

"They said 'fine, do it however you want; we'll do what you tell us to do.' So I started leading them."

"How did you conduct the prayer?"

"We gathered in a circle, and I stood in the middle, and I led them in prayer, and we gave ourselves over to God. We started right away, because we didn't have any other way out."

"Did the whole group participate in the prayer sessions?"

"The whole group, all thirty-three."

"Irrespective of everyone's particular religious background?"

"Yes, because we all wanted the same thing: to put ourselves in God's hands, because he was the only one who could find a way out and guide the people up on the ground to where we were so they could rescue us."

"What was your first prayer?"

"We prayed for the rescue workers to be able to get through the main tunnel so they would find us—that was the first thing—but then we accepted that it was impossible because there was a huge boulder blocking the way in from there."

"But you were all sure that they were looking for you?"

"Of course. We knew our families wouldn't just let us go; they would never do that."

"Did you pray on a set schedule, or just anytime?"

"We had two prayer sessions each day: one at noon and the other at six in the evening. We went around the circle and prayed for the same things, and we started talking about the word of God, of the messages he sent. Since we didn't have a Bible, I told them about what I knew."

"What was the group dynamic like? Were you always in charge of leading the sessions?"

"No, I started to give the others a chance, because that was what it was about—that it wasn't just me sitting pretty in the middle—and I started asking the others to speak up. So it started off with me praying, and then asking some of the others to pray, and then it turned into a whole service, with call and response, and passages from the Bible, and with a moment of silent prayer, too."

"Was everybody enthusiastic about it?"

"There were a few who didn't feel comfortable praying just like everyone else, but they did in their own way. They prayed in their own way. They were praying just like everyone else in their own words, they way they wanted to."

"So the group had faith in you as a spiritual leader?"

"Although I'm not an official pastor or spiritual guide, I am a responsible person. I mean, I live in truth; I know that God is a living God; I know because he is present in my life. He is a God who makes himself known. That's why I told them that when you give yourself over to God, he always, always becomes a living presence in your life."

José Henríquez remembers that after they made contact with the people on the surface and had some Bibles sent down to them, toward the end of their time trapped in the mine they started a kind of basic Bible study.

"It was like a course, with questions and answers. A lot of the men participated, and whoever got the most answers correct on a quiz with thirty questions won a Bible. I gave them a whole afternoon to study the Bible before taking the quiz."

"Who won?"

"Ariel Ticona did. He answered all thirty questions right. He was so sure he would win, on his answer sheet he wrote in the margin, 'The Bible is mine!'"

"How do you think God manifested himself to all of you?"

"He didn't have to blast through the mine to be by our side. It was the Holy Spirit that moved us, that blessed us, that gave us strength, the inner strength we needed to be all right, that was the spirit that touched the hearts of us thirty-three miners through the word of God. That is what I can testify, that God was there with us." José assures me that the success of the drilling that was undertaken to locate them alive was nothing short of a miracle.

"What's important to me is to testify that it was a miracle. The experts can't explain how that drill could have gone so many meters off its trajectory so it could then run into us. Don't forget that on the surface they were drilling blind. They weren't going by topography or anything. They were trying to run into any hollow or tunnel, to somehow come across us however they could. That was what we wanted: for one of those boreholes to come close enough that they could hear us and we could send a message."

Excited now, Henríquez recalls how they prepared for the drill to reach them, and how overjoyed they were when it broke through the earth.

"We knew they were trying to get to us through the boreholes. Drilling is the first thing they would do to find out what happened in that kind of situation. We had our messages ready to send up to the surface. We had paint to mark the end of the drill—several cans of spray paint that I normally used on the machinery in my job. And we found lots and lots of wire to attach the notes to the drill."

"And what did your message say?"

"I didn't have a message; I just had the paint. I said, 'Here's some paint; that's my contribution. And I found a big metal rod to hit the end of the drill with, to make a sound that they could hear from the surface, so they'd go, 'Hey there are people down there;

they're banging on this—listen!' The idea was to do whatever we possibly could to let them know we were alive."

"How did you all react when the drill broke through?"

"First we started to hear some noise, getting louder and louder, until the drill broke through the rock, and it was in the tunnel. When it appeared, a lot of the guys tried to hug it, out of sheer happiness!"

José Henríquez shifts in his chair. After a pause, he continues. "The drill stopped for a while. So we hurried to mark it with spray paint. We wanted to just grab on to the drill and let it pull us all back up with it, but we couldn't do that. We were really happy— really, really happy."

"So, then, there wasn't any doubt that you would be rescued?"

"No doubt. A window of salvation had been opened for us, and we knew God was going to bless us through that drill and give us a way to communicate with the surface, and that they could send fresh water and food down to us. I think all of that was a clear sign that we would be rescued." Closing his eyes, immersed in his un-shakeable faith, José says softly, "It was another sign from God . . . one more sign."

"Up until that point, had the conditions in the refuge deteriorated?"

"Yes, sure. By then we had really had it. We couldn't have lasted much longer. We were really in trouble, from the hunger, the basic issues."

"What precautions did you take to make sure the drilling didn't do any damage—like a rock hitting one of you on the head, for example?"

"We had to be careful and gather in an area where there wasn't as much vibration and dust. Of course we didn't want them to stop drilling, but there was a lot of dust flying around from it."

"There was no way to make the dust dissipate?"

"Not at first, no. Then they poured water down over the drill from above, and that helped the dust to settle, and we had to yell and scream because our refuge was piling up with dirt and debris."

Finally, José's wife prevails, and our conversation comes to an end. He has to attend a family function, although he will be a bit late. Before leaving, he glances at his watch in surprise: the ten minutes he had initially slated for our talk had run out long ago.

CHAPTER SEVENTEEN

SON OF
THE DESERT

The almost nonexistent rain in the Atacama Desert results in
a strange landscape for the mining camps of Norte Grande
de Chile. The houses have flat roofs—just horizontal plat-
forms. The children think nothing of climbing up to retrieve a ball
lost through an errant kick during *"pichangas,"* the informal soccer
games that fill the afternoons, going on until the sun sinks slowly on
the horizon, firing the mountains with colors from dazzling orange
to soft purple and sending the players home.

The architecture is perfectly uniform. The family shelters, of an
ocher that blends with the color of the hills, have little small yards
with some semblance of gardens, cactuses, little plots of pasture,
peppers, and any other kind of plant that doesn't need watering.
Water is strictly for human consumption. What the Atacama has
in abundance is the clean immensity of its skies, the clearest on the
planet. And beneath them, the earth dominates every trait and aspect of its inhabitants. The dryness, day by day, cracks the skin and

hands but also strengthens the spirit of those who live by what the mineral veins promise.

Meanwhile, the soporific dust, borne on the blustery north wind, steals in through the small spaces around the doors and windows of Víctor Zamora's house. Here he spends the days reconstructing the events of his imprisonment for more than two months, nearly half a mile underground. A confinement that changed his life and the lives of his comrades, just as it disrupted the life of all Chile and sent a shiver through millions all over the world.

He looks with disdain at the glass of water beside him and refuses it, almost annoyed, even though his throat gets drier with every word he speaks. True, it isn't easy to gain the trust of a Chilean miner. They are hard men, matching the hard earth of the Atacama.

"I thought I would never get out of that mine," he says. "I tossed and turned in the refuge, remembering my son, my wife . . ."

Zamora fidgets uncomfortably in his chair. He seems uneasy and can't find the right position to sit in. It seems as if his soul doesn't yet fit in this rebirth above the ground, beneath the sun. It's impossible not to look at the tattoos adorning the sinewy hardness of his arms.

"Each one represents a stage in my life," he says.

As if in a gallery of miniatures, the face of another Víctor emerges on Zamora's arm. It is Víctor Jara, the Chilean singer murdered by soldiers after the coup d'état led by Augusto Pinochet in 1973. And on Zamora's left arm is Che Guevara, with his trademark beret, the image blurred by age.

Farther down, popping up almost whimsically, are the words "I love you," the face of Lucifer, and a marijuana leaf, which, Víctor explains, recalls the more psychedelic stage of his life.

After his experience in the San José mine, he says he is preparing a tattoo for his back, though he isn't ready to reveal the design. It could have mystical or divine origins, but that is among the secrets that Zamora holds from his sojourn in hell—secrets perhaps bigger even than he imagines.

Víctor was born into a family of miners, accustomed to frost at dawn, the blast furnace of the midday sun, the chill of sunset, the long shifts without a rest break, the closed and inhospitable houses, the warmth of loved ones, the alcohol with pals, the endless sacrifice, the pride in being a miner.

He grew up this way, he has suffered and lived this way, and this way he will continue, while hoping—though with doubts—that his children might find an easier path.

The day of the rescue, Víctor was the fourteenth survivor to breathe the clean night air of freedom. A few steps away waited his four-year-old son, Arturo, and his wife, with a new life of three months growing inside her.

The long-awaited embrace lasted an eternity for them, though only seconds for the millions seeing the moment of their reunion live on television.

A TRAGEDY FORETOLD

"The day of the accident, something very strange happened," Víctor Zamora recalls. "I received a warning, a premonition, that something bad could happen to me. My little Arturito has always crossed me when he gets home from school and I have to leave for work. He leaves school at four and I start work at six; so he would give me a kiss on the lips, cross me, and say, 'Have a good one, and may God watch over you.' That day, he didn't do it; he just didn't feel like it. It seemed strange to me, because he's really attached to me. And I told him I would bring him his snack, and he told me that the snack didn't matter, that he wanted me to stay, not to go to work. . . . I didn't pay him any mind."

It is three in the afternoon. The heat is blasting, and the howl of the wind now and then is the only thing to interrupt the silence. It's a silence that commands respect.

"So you would fix his lunch?"

"No, at the company they gave me juice, cookies, milk, so instead of eating those things, I would save them for my son so he could take them to the school."

"Even though you would be hungry?"

"Yes, anything for him. Also, that morning I felt someone touch my shoulder. I turned to look, and no one was there."

"How's that?"

"In the morning, I got up around five thirty. I had just gone to the kitchen to heat a little water for tea. I felt bad; my stomach was bothering me. I was standing with my jacket in my hand, by an armchair, and it was as if someone grabbed me by the shoulder and tried to shove me down into the chair. I looked all around, and nobody was there. I was alone. I laughed at myself, but later it scared me. All day I had a knot in my throat."

He clears his dry throat and looks again at the glass of water, indifferent. With an effort, he goes back to remembering.

"What were you doing the day the mine collapsed?"

"We always went out at two P.M. to eat lunch. Right then there were maybe twelve guys in my group. Then a siren rang, and the cart where I was started having trouble moving forward. The opening of the mine started to crumble, rocks were coming down. We were a hundred and ninety meters deep. I was with Carlos Barrios, and there were two more: José Ojeda and Claudio Acuña. We backed up a little and it felt like the earth was falling in on all sides. There was a tremendous noise, a huge wave of pressure. We couldn't continue upward, so we went down. There was dust everywhere. Then it calmed down a little. So my mates went to look for pickups, and we got in the one that Florencio Ávalos always drove, and tried again to get out, but it was impossible. The exit was plugged by this monster of a rock. That's when I realized we were trapped, and I felt that I would never get out of that place."

Finally, he takes a sip of water, and even though it's tepid by now, it still lets him continue talking.

"How did you all react?"

"Well, for my part, I believed this was it for me, that I was never getting out . . . ever. My mates, I don't know what they thought; I don't know what their experience was in that moment. Then we all went to the refuge to wait till the dust cleared. There was barely any light, because the little lamps we had were going dead.

"So going to the refuge was the first decision?"

"Of course. We went down to see what we were going to do, see what food there was."

"What did you find?"

"Something humiliating for a working man. There was hardly any food, and maybe a hundred or so sets of silverware. I don't understand—for what? We looked at each other and didn't know how we could make it with so little food. There was milk, but it was curdled, and there was food for two days, but after that we wouldn't have anything to eat. Also, there was a can of peaches and two cans of peas, and something like eight liters of good milk and some more that was bad—about half and half—and there were maybe nineteen cans of tuna. One of my mates has the list of all that. I think there were something like fifteen liters of water. So there we were in a totally humiliating situation. Imagine: the owners of the mine make all that money, and to have something like that! Go figure," he muses, shaking his head. He pats his hair and impulsively grabs his chin.

"How did you all plan on getting out after that?"

"I think we all accepted what was happening. That was our common ground: Accept that we were trapped. But also we had to accept that we had to get out of that hole alive, and for that we needed to make the food last."

After not moving for quite a while, Víctor shuffles in his chair. He rubs his eyes before reentering a difficult spot in the tragedy.

"What did you do those first days? What was your routine like?"

"We would sleep an hour, an hour and a half, and wake up. Then we'd try to sleep again; I think in twenty-four hours we slept about three. The rest of the time we waited for some light, some hope, something . . ."

"How did you arrange for sleeping?"

"Right there on the rock, on that damp ground, with our feet like so. You just had to make yourself comfortable as you could—throw the rocks over there and use what little clothes you had to make some kind of little pillow."

"How did you, um, take care of sanitary needs?"

"At first there was no need—I guess nobody felt the urge. I couldn't take a leak. Some of the guys found a little corner a little ways off to piss. As far as the other, nothing . . . since our bellies were empty."

"How did you handle decision making?"

"Talking, giving opinions about this and that. Everything went up for discussion, and whatever got the most votes won. Or rather, everybody had the right to talk. Now, of course some talked more than others, but everybody said something, although sometimes we rushed it because a lot of us were talking at the same time, anxious to say something. The more good ideas, the better for the whole group, I say."

"If you didn't have much food, how did you divide it?"

"We looked, and it was just about nothing. So we calculated the food for ten days—eating the minimum, of course. At that point, we decided, every twelve hours, to eat a teaspoonful of tuna each, then every twenty-four, thirty-six, and later on every forty-eight hours. Nobody got more than anybody else—everybody the same. We had such hunger, I'd just as soon give it to you instead. . . . Every day that went by, we ate less and less. We were getting to the point where we went seventy-two hours without a bite, when we had only one can of tuna left."

"You shared it?"

"No, that one we didn't open. We left it in the mine as a symbol of our survival."

Then, in halting sentences, his gaze switching back and forth, eyes shiny with tears, Víctor Zamora reveals the greatest pain that he suffered during his confinement underground.

"The worst thing I felt wasn't the hunger, or the sad time we spent buried in the mine, but the fact of not being able to see my family, to see my boy, to see my wife smile, to see my mom, my brothers and sisters, my grandma . . . I think the idea of not seeing them ever again just killed me. That was the worst agony, the most awful pain I have ever felt. I don't know about my mates, but I believe they'll say pretty much the same thing. In those moments . . . those most critical moments, one really . . ."

He pauses, looks up, breathes deeply, and finds again the voice of a strong man before continuing. "We really didn't have anything. We didn't take in anything but water, I believe every twelve hours we waited to eat something, but I had an ache in my soul from not being able to see my son, to hug him. I was attached to my family from way down in the depths of the earth, so now that I'm out, I try to feel that again, but it's getting a little tough for me."

"Why?"

"Because I keep going back to it, to those memories I have stuck in my head. So I crave silence. I need to write and I'm going to keep on writing . . . it's an urgent need."

THE FIRST DAYS

"Although we were in darkness, we also knew whether it was night or day, because one of us had a watch to guide us. The most important thing is that those first days we tried to send signals to the surface. We made a fire so the smoke would go up the ventilation duct, but nothing got through. We were desperate. One of my mates shouted and shouted for help, but what were they going to hear up there from us, seven hundred meters down?"

With the pitiless passing of the days, the spirit of these tough men began to crumble. The first symptoms of depression started to show. Sadness and sorrow gradually took over the group, although no one dared to be the first to acknowledge it.

In the refuge of the San José mine, anxiety was a sign of a weakness that the men did not inherit from their parents, that they never wanted their children or their wives to experience. They were strong men; they were pillars of their families—until life changed. It was in that moment when the figure of José Henríquez began to shine. A miner of endless struggles, Henríquez, thanks to his intimate closeness to the Bible, found his natural role—as spiritual leader of the group.

"At first, we needed words to give us spirit. Then the idea occurred to us to call on don José. He was a very quiet guy, didn't talk a lot, and with him being on my crew, I knew he went to the church. The decision was made. He could give us, anytime, a word or two of encouragement—having to do with God, of course—and that's how it all started."

Quite out of the blue, Víctor Zamora stops his account with a silence that threatens to be final. He seems to gaze off into the emptiness, an emptiness that seems to live inside him, as if he were obliged to look for another path among his memories, his nightmares, and his hope.

Again he grows still, absent, perhaps buried once more beneath that near half-mile of doubt. The seconds stack up until they seem to fill the room, until, eyes brimming with tears, Víctor raises his head and returns to the here and now. His gaze is like that of a guilty child who has been scolded harshly by its mother. Inexplicably, he apologizes and then exhales, stringing together syllables that join to form words.

"Around noon, we all would pray, ten minutes, fifteen minutes, and something happened . . . I don't know, as if something gave us a strength of will to keep fighting during those critical moments that we were going through. And as the days went by, it became a habit. The refuge turned into a little temple. I took it that way, and then, with each passing day, when our strength flagged and when don José spoke the word, a strange force went out over us, enabling us to stand, and to stay standing, for the daily prayer."

Víctor says that living a step away from death obliged them to take refuge in a higher being, the only one capable of saving them.

"Look, that decision was somehow more personal, it was of that moment. I believe that we all needed a guiding light, an opportunity to hold on to something. So what better to hang on to than God? The best instrument we had was the word, and so it was that we all began, bit by bit, to love him and ask him for things and, at the same time, to repent for everything wrong we had done, and to ask for what we needed in that moment, which was to get out, get out, get out . . . to be alive."

Without so much as an effort to smile, Zamora reveals his conviction that the experience in the mine signified a rebirth that allowed him to straighten his path from the wild parties of his miner's world, which he indulged in from time to time before the accident.

"Well, I am very grateful to God for giving me a new life, to be born again, and as I said before, I believe in God in my own way. I attest that he exists, and I take that as a fact—yes, as a fact. Now I go everywhere with him, I talk with him, and I ask him for all good things for my life, for my children, my wife, my family—everything a man ever wanted."

Víctor Zamora pauses. He crosses his hands and lays them on the table. For several seconds, he doesn't move a muscle. Again he breathes deeply, forcefully—before making an unexpected and intimate confession.

"I'm going to tell you something. I wrote a letter to my mother when I was down there in the mine and in a bad way, after reading something in the Bible. I don't know what made me open it. I had opened the Bible many times, but I never understood it. But one sentence stuck in my head: 'if you don't love your brother, you will not enter the kingdom of God.' Wow. That gave me something to think about."

His worry shows. He starts to speak, stops, tries to find the right words to express what he feels.

"I told my mom that in all our family we have to be good to each other. I asked her if she would show that verse to my brothers, because I wanted them to understand that we shouldn't wait until one of us is in a bad situation or someone dies for the family to come together, but rather, we should be good to each other before circumstances make us."

Víctor says he wrote this letter to bring his family together. After surviving hell, he seems to have taken this on as his new challenge.

"That occurred to me in those moments when we started to pray as a group. There were Catholics, Evangelicals, Jehova's Witnesses, but inside there was just one church, and we all prayed for the same thing, all loved the same God, and that was good. I stressed to my mother how beautiful that meeting was . . . and when her reply came back to me . . ."

At this point all is silent.

For the first time, there is the hint of a smile, and all he manages to say is, "It was a nice experience."

"What did your mother say?"

"That the words I had expressed in the letter were a message from God and that now I had better change—be another person, with more maturity, and convey faith and hope to them, to my brothers, that I was going to leave the mine with a gift. Sadly, the letter I wrote my mother seems to have gotten lost. She's looked high and low for it, but she can't find it."

Víctor says his mother's reply moved him to keep writing there in the mine, to be useful with his poetry.

"Instead of weighing down my spirit with all that negative talk about my sorrows, I tried to convey my messages to the others. I wrote mainly to give thanks. That's why I dedicated myself to writing. It was maybe the fourth time we were praying together that I got the feeling to tell everyone what we were really living, experiencing, and I hope everyone in the world realizes that they don't have to wait for some accident or some misfortune before they ask forgiveness or, for example, approach someone they've stopped talking to . . . be friendly toward them. I was inspired in my pain, in the suffering of my mates. I wanted to lift their spirits. We had so many problems, but I absorbed the pain to give them courage. I saw such pain."

Víctor promises that one day he will have his letters published. He is absolutely convinced that those stories will help many others relieve the afflictions of their soul, of every soul, and draw those same life lessons, starting with what he experienced in the bottom of the mine.

"So they are going to be able to see for real the pain I felt in that moment for my comrades, for myself, for my family, because I got it across with a poem. I was empathizing with the suffering of others, with the anguish of all my fellow miners."

NOTHING SO BITTER

Víctor Zamora was recognized by his coworkers as someone who was always ready with a joke. And he often used that ability to lift the spirits of his overwhelmed coworkers. Sometimes they laughed at his jokes, sometimes at him.

"At certain moments I tried to laugh about our misfortune, cracking jokes at my mates . . . Some were a little cruel," he recalls.

He laughs for a couple of seconds and takes the moment to relax.

"One day, when we were playing dominoes, I suddenly started talking about food: tasty dishes, hot dogs, barbecue. . . . Everybody was ready to kill me. But they paid me back in kind."

"How?"

Zamora laughs. "I can't tell you. . . . But everything was just to lift our spirits with jokes, even though many were rude or inappropriate."

"Were there times when those jokes didn't go over well?"

"Yes, that happened afterward, that is, when we were eating the food they sent us from up above. I believe that when a person regains his strength and is back in shape again, they start to be the same as they always were. Before, we were united by distress, fear. The thirty-three were just one. After they got to us with the drill that punched into the refuge, we all went back somehow to being the same people we were before. Personal differences came up. There were arguments, for sure, but we never came to blows. We cursed each other plenty of times, but we never went to extremes. . . . There were words but never real clashes."

"Did you insult each other much?"

"Mmmm! For me, getting cussed at is no big deal; it's part of our way of speaking. For me, it's only a fight if someone touches you, but that never happened among us."

"This may be a dumb question, but what motive did people have for fighting?"

"Let me see if I can explain. There are people who get mad over whatever little thing, or suddenly the person is in a bad humor, and that's enough to start it, and I found that normal there in the mine. There were loud arguments between my mates, but a half hour later they were apologizing and talking about it like it was nothing. Then they would share a maté, tea, a cup of coffee, and that was it—no big thing."

"But when the food started coming and the outlook improved, everyone's behavior changed?"

"That explanation is a little too simple. We were at the point of dying, so the only way to be together was to unite like brothers. So in critical moments, we were all fighting for the same thing: to stay alive. After the drill broke through and food started coming and we started regaining our strength, then the thing each person missed was their family. I believe that was why the quarrels erupted: out of frustration. Listen, because everyone was fed—and I include myself—we changed our way of speaking to each other. The misfortune, the time of sleeping in the same bed with death—that was getting further and further behind us. I felt that I would live, but I was still tucked away in the center of the earth, not with the ones I love. I remember that we could see our families through a camera that they lowered to us, but it was tremendously frustrating. Then, seeing my son, it made me cry—the same with my wife, because I saw the images. I asked myself, *Why can't I be out there? Why can't I see the sun?* On top of everything, the authorities up above kept a protocol that we down below thought was absurd."

"Really so bad?"

"No, it was a rule of theirs, like a routine, saying, 'today you're having coffee and toast,' and then you had to agree to that. And there was the rule that my family couldn't express any negative feelings, so that I wouldn't cry, so that I wouldn't feel bad. And I couldn't say anything to make them feel bad, either; so we felt like we were going to explode, and that was the last straw. People started having problems with each other. They were little things, sure, but it's just that the human being is used to expressing itself whether things are good or bad."

PHYSICAL PAIN

"I was really sick. I broke my teeth, and my implant came out, and I ended up with an infected molar. I had a little cavity, and down there, with the lack of calcium, and all the things going on, I lost my real tooth and the implant. In any case, I didn't worry too much about my teeth, because the situation we were in, there wasn't time for physical pain. To forget about my situation a little, sometimes I would sit and play dominoes, which Luis Urzúa made out of some plastic he found lying around. We cut it into squares and then made the little holes with a lighter. We played just for fun. Whoever wanted to play could play. And that's how we passed the days until the drill appeared."

"What did you all do when the drill broke through the rock and came into the refuge?"

Víctor smiles again—the third or fourth time in several hours—as the wind outside slacks off and the chill of sunset steals over the Atacama. Lifting his hand, he strokes his cheek and seems to be looking for the words.

"It was good, it was really good, spectacular. It was as if someone reached down an arm with all the strength in the world, and there we were. How we all cried! It seems that . . . [He sobs]"

Víctor clearly cannot keep talking. His mouth trembles, and he asks if we can end the conversation.

Víctor has been strong, throughout the ordeal and this interview—imposing order on so many jumbled thoughts, quieting the voices of the nightmares, revisiting and prolonging the emotions still tangled up between mind and spirit, between trauma and hope, between the shared agony and his new life.

His second life.

EPILOGUE

The rescue of the thirty-three trapped miners at the San José mine left us with many lessons, as varied as they are controversial. From an organizational and corporate perspective, the experience can serve as a shining example of successful crisis management. The authorities firmly assumed control of the situation and confronted the dangers head-on. They understood that time was a critical factor, and they were able to marshal the technical know-how and labor necessary to get the job done. If anything went as well as it could have right after the mine's collapse, it was the rapid, correct assessment and acceptance of the situation, and the immediate formulation of a rescue plan.

From a political perspective, the right has asserted that the story's happy ending can be attributed to the triumph of capitalism. The machinery and technology used in the rescue, the argument goes, were built and transported to Chile by foreign business interests whose main objective was making a profit. For example, the cable that was attached to the Phoenix capsule came from Germany. Japan supplied the fiber optic cables that allowed the miners to communicate with the rest of the world. Samsung, a South Korean company, sent video telephones so the miners, trapped almost half a mile below the earth, could communicate live with their families above, and their relatives could see them. Clearly, the free market itself was the winner here, they reason.

Also, standardized labor protocols that came into play during the rescue efforts were borrowed from the most successful practices of private industry. If the accident had taken place in a country with a closed economy, like Venezuela or Cuba, would the miners have had the same luck? Political conservatives wonder.

But other voices more critical of Chilean society are calling for a deeper examination of the causes of the accident and its consequences. At the heart of that discussion must be warnings about the unsafe working conditions that have historically been imposed on miners. On top of that, regulatory measures for healthcare and safety standards in the workplace have been relaxed. After the San José rescue, what has hardly even been mentioned in the public discourse are the other mining accidents and deaths that have occurred in the mines in northern Chile since August 5, 2010, because of unsafe working conditions.

A few weeks after the last of the miners was lifted by capsule to safety, another young miner died in the same region of Atacama from unsafely handling explosives. The victim was identified as Héctor Mauricio Cortes Vergara, and the two men wounded in the accident were Juan Carlos Andres Lara and Juan Carlos Duran Castillo; the latter suffered a very serious injury.

On November 9, 2010, a similar event at the Los Reyes mine, in the northeast of Copiapo, resulted in two deaths and one serious injury.

Another miner died on November 30, 2010, at the Tambillo mine in the Coquimbo region in northern Chile. He was pulled into a conveyor belt and crushed. Only twenty-four hours later, another man lost his life at the Pelambres mine in the same region, as he performed maintenance work on heavy machinery.

Neither the Minister of Mining Laurence Golborne or President Piñera have traveled to those mines to investigate or condemn the tragedies.

Now, in view of all that has happened, Chile's working classes and labor organizations must seize on the public's outpouring of sympathy that came out of the rescue operation and push for the changes needed to ensure that what happened to the thirty-three miners at San José, and to the men in the other more recent disasters, never happens again.

INDEX